Graduate Texts in Mathematics 7

Springer
New York
Berlin
Heidelberg
Barcelona
Hong Kong
London
Milan
Paris
Singapore
Tokyo

Graduate Texts in Mathematics

(continued after index)

J.-P. Serre

A Course in Arithmetic

Jean-Pierre Serre
Collège de France
75231 Paris Cedex 05
France

Mathematics Subject Classification: 11-01

Title of the French original edition: *Cours d'Arithmétique*.
Publisher: Presses Universitaires de France, Paris, 1970–1977.

Library of Congress Cataloging in Publication Data
Serre, Jean-Pierre.
A course in arithmetic by J.-P. Serre. New York,
Springer-Verlag 1973

viii, 115 p. illus. 25 cm. (Graduate texts in mathematics, 7)
Translation of Cours d'arithmétique.
Bibliography: p. 112–113.

1. Forms, Quadratic. 2. Analytic functions.
I. Title. II. Series.
QA243.S4713 512.9'44 70-190089
ISBN 0-387-90040-3; 0-387-90041-1
(pok.) MARC

Printed on acid-free paper.

Printed and bound by R. R. Donnelley and Sons, Harrisonburg, VA

9 8 7 6

ISBN 0-387-90040-3
ISBN 3-540-90040-3 SPIN 10783650

Springer-Verlag New York Berlin Heidelberg
A member of BertelsmannSpringer Science+Business Media GmbH

Preface

This book is divided into two parts.

The first one is purely algebraic. Its objective is the classification of quadratic forms over the field of rational numbers (Hasse-Minkowski theorem). It is achieved in Chapter IV. The first three chapters contain some preliminaries: quadratic reciprocity law, p-adic fields, Hilbert symbols. Chapter V applies the preceding results to integral quadratic forms of discriminant ± 1. These forms occur in various questions: modular functions, differential topology, finite groups.

The second part (Chapters VI and VII) uses "analytic" methods (holomorphic functions). Chapter VI gives the proof of the "theorem on arithmetic progressions" due to Dirichlet; this theorem is used at a critical point in the first part (Chapter III, no. 2.2). Chapter VII deals with modular forms, and in particular, with theta functions. Some of the quadratic forms of Chapter V reappear here.

The two parts correspond to lectures given in 1962 and 1964 to second year students at the Ecole Normale Supérieure. A redaction of these lectures in the form of duplicated notes, was made by J.-J. Sansuc (Chapters I–IV) and J.-P. Ramis and G. Ruget (Chapters VI–VII). They were very useful to me; I extend here my gratitude to their authors.

J.-P. Serre

Table of Contents

A Course in Arithmetic

Part I

Algebraic Methods

Chapter I

Finite Fields

All fields considered below are supposed commutative.

§1. *Generalities*

1.1. *Finite fields*

Let K be a field. The image of $\mathbf{Z}$ in K is an integral domain, hence isomorphic to $\mathbf{Z}$ or to $\mathbf{Z}/p\mathbf{Z}$, where p is prime; its field of fractions is isomorphic to $\mathbf{Q}$ or to $\mathbf{Z}/p\mathbf{Z} = \mathbf{F}_p$. In the first case, one says that K is of *characteristic zero*; in the second case, that K is of *characteristic* p.

The characteristic of K is denoted by $\operatorname{char}(K)$. If $\operatorname{char}(K) = p \neq 0$, p is also the smallest integer $n>0$ such that $n.1 = 0$.

Lemma.—*If* $\operatorname{char}(K) = p$, *the map* $\sigma: x \mapsto x^p$ *is an isomorphism of* K *onto one of its subfields* K^p.

We have $\sigma(xy) = \sigma(x)\sigma(y)$. Moreover, the binomial coefficient $\binom{p}{k}$ is congruent to 0 (mod p) if $0<k<p$. From this it follows that

$$\sigma(x+y) = \sigma(x)+\sigma(y);$$

hence σ is a homomorphism. Furthermore, σ is clearly injective.

Theorem 1.—i) *The characteristic of a finite field* K *is a prime number* $p \neq 0$; *if* $f = [K:\mathbf{F}_p]$, *the number of elements of* K *is* $q = p^f$.

ii) *Let* p *be a prime number and let* $q = p^f (f \geqq 1)$ *be a power of* p. *Let* Ω *be an algebraically closed field of characteristic* p. *There exists a unique subfield* $\mathbf{F}_q$ *of* Ω *which has* q *elements. It is the set of roots of the polynomial* $X^q - X$.

iii) *All finite fields with* $q = p^f$ *elements are isomorphic to* $\mathbf{F}_q$.

If K is finite, it does not contain the field $\mathbf{Q}$. Hence its characteristic is a prime number p. If f is the degree of the extension $K/\mathbf{F}_p$, it is clear that $\operatorname{Card}(K) = p^f$, and i) follows.

On the other hand, if Ω is algebraically closed of characteristic p, the above lemma shows that the map $x \mapsto x^q$ (where $q = p^f$, $f \geqq 1$) is an automorphism of Ω; indeed, this map is the f-th iterate of the automorphism $\sigma: x \mapsto x^p$ (note that σ is surjective since Ω is algebraically closed). Therefore, the elements $x \in \Omega$ invariant by $x \mapsto x^q$ form a subfield $\mathbf{F}_q$ of Ω. The derivative of the polynomial $X^q - X$ is

$$qX^{q-1} - 1 = p.p^{f-1}X^{q-1} - 1 = -1$$

and is not zero. This implies (since Ω is algebraically closed) that $X^q - X$ has q distinct roots, hence $\mathrm{Card}(\mathbf{F}_q) = q$. Conversely, if K is a subfield of Ω with q elements, the multiplicative group K^* of nonzero elements in K has $q-1$ elements. Then $x^{q-1} = 1$ if $x \in K^*$ and $x^q = x$ if $x \in K$. This proves that K is contained in $\mathbf{F}_q$. Since $\mathrm{Card}(K) = \mathrm{Card}(\mathbf{F}_q)$ we have $K = \mathbf{F}_q$ which completes the proof of ii).

Assertion iii) follows from ii) and from the fact that all fields with p^f elements can be embedded in Ω since Ω is algebraically closed.

1.2. *The multiplicative group of a finite field*

Let p be a prime number, let f be an integer $\geqq 1$, and let $q = p^f$.

Theorem 2.—*The multiplicative group $\mathbf{F}_q^*$ of a finite field $\mathbf{F}_q$ is cyclic of order $q-1$.*

Proof. If d is an integer $\geqq 1$, recall that $\phi(d)$ denotes the *Euler ϕ-function,* i.e. the number of integers x with $1 \leqq x \leqq d$ which are prime to d (in other words, whose image in $\mathbf{Z}/d\mathbf{Z}$ is a generator of this group). It is clear that the number of generators of a cyclic group of order d is $\phi(d)$.

Lemma 1.—*If n is an integer $\geqq 1$, then $n = \sum_{d|n} \phi(d)$.* (Recall that the notation $d|n$ means that d divides n).

If d divides n, let C_d be the unique subgroup of $\mathbf{Z}/n\mathbf{Z}$ of order d, and let Φ_d be the set of generators of C_d. Since all elements of $\mathbf{Z}/n\mathbf{Z}$ generate one of the C_d, the group $\mathbf{Z}/n\mathbf{Z}$ is the disjoint union of the Φ_d and we have

$$n = \mathrm{Card}(\mathbf{Z}/n\mathbf{Z}) = \sum_{d|n} \mathrm{Card}(\Phi_d) = \sum_{d|n} \phi(d).$$

Lemma 2.—*Let H be a finite group of order n. Suppose that, for all divisors d of n, the set of $x \in H$ such that $x^d = 1$ has at most d elements. Then H is cyclic.*

Let d be a divisor of n. If there exists $x \in H$ of order d, the subgroup $(x) = \{1, x, \ldots, x^{d-1}\}$ generated by x is cyclic of order d; in view of the hypothesis, all elements $y \in H$ such that $y^d = 1$ belong to (x). In particular, all elements of H of order d are generators of (x) and these are in number $\phi(d)$. Hence, the number of elements of H of order d is 0 or $\phi(d)$. If it were zero for a value of d, the formula $n = \sum_{d|n} \phi(d)$ would show that the number of elements in H is $<n$, contrary to hypothesis. In particular, there exists an element $x \in H$ of order n and H coincides with the cyclic group (x).

Theorem 2 follows from lemma 2 applied to $H = \mathbf{F}_q^*$ and $n = q-1$; it is indeed obvious that the equation $x^d = 1$, which has degree d, has at most d solutions in $\mathbf{F}_q$.

Remark. The above proof shows more generally that all finite subgroups of the multiplicative group of a field are cyclic.

§2. *Equations over a finite field*

Let q be a power of a prime number p, and let K be a field with q elements.

2.1. *Power sums*

Lemma.—*Let u be an integer $\geqq 0$. The sum $S(X^u) = \sum_{x \in K} x^u$ is equal to -1 if u is $\geqq 1$ and divisible by $q-1$; it is equal to 0 otherwise.*

(We agree that $x^u = 1$ if $u = 0$ *even if* $x = 0$.)

If $u = 0$, all the terms of the sum are equal to 1; hence $S(X^u) = q.1 = 0$ because K is of characteristic p.

If u is $\geqq 1$ and divisible by $q-1$, we have $0^u = 0$ and $x^u = 1$ if $x \neq 0$. Hence $S(X^u) = (q-1).1 = -1$.

Finally, if u is $\geqq 1$ and not divisible by $q-1$, the fact that K^* is cyclic of order $q-1$(th. 2) shows that there exists $y \in K^*$ such that $y^u \neq 1$. One has:

$$S(X^u) = \sum_{x \in K^*} x^u = \sum_{x \in K^*} y^u x^u = y^u S(X^u)$$

and $(1-y^u)S(X^u) = 0$ which implies that $S(X^u) = 0$.

(*Variant*—Use the fact that, if $d \geqq 2$ is prime to p, the sum of the d-th roots of unity is zero.)

2.2. *Chevalley theorem*

Theorem 3 (Chevalley–Warning).—*Let $f_\alpha \in K[X_1, \ldots, X_n]$ be polynomials in n variables such that $\sum_\alpha \deg f_\alpha < n$, and let V be the set of their common zeros in K^n. One has*

$$\mathrm{Card}(V) \equiv 0 \pmod{p}.$$

Put $P = \prod_\alpha (1 - f_\alpha^{q-1})$ and let $x \in K^n$. If $x \in V$, all the $f_\alpha(x)$ are zero and $P(x) = 1$; if $x \notin V$, one of the $f_\alpha(x)$ is nonzero and $f_\alpha(x)^{q-1} = 1$, hence $P(x) = 0$. Thus P is the *characteristic function* of V. If, for every polynomial f, we put $S(f) = \sum_{x \in K^n} f(x)$, we have

$$\mathrm{Card}(V) \equiv S(P) \pmod{p}$$

and we are reduced to showing that $S(P) = 0$.

Now the hypothesis $\sum_\alpha \deg f_\alpha < n$ implies that $\deg P < n(q-1)$; thus P is a linear combination of monomials $X^u = X_1^{u_1} \ldots X_n^{u_n}$ with $\sum u_i < n(q-1)$. It suffices to prove that, for such a monomial X^u, we have $S(X^u) = 0$, and this follows from the lemma since at least one u_i is $< q-1$.

Corollary 1.— *If $\sum_\alpha \deg f_\alpha < n$ and if the f_α have no constant term, then the f_α have a nontrivial common zero.*

Indeed, if V were reduced to $\{0\}$, $\mathrm{Card}(V)$ would not be divisible by p.

Corollary 1 applies notably when the f_α are *homogeneous*. In particular:

Corollary 2.—*All quadratic forms in at least 3 variables over K have a non trivial zero.*

(In geometric language: every conic over a finite field has a rational point.)

§3. *Quadratic reciprocity law*

3.1. *Squares in* $\mathbf{F}_q$

Let q be a power of a prime number p.

Theorem 4.—(a) *If* $p = 2$, *then all elements of* $\mathbf{F}_q$ *are squares.*

(b) *If* $p \neq 2$, *then the squares of* $\mathbf{F}_q^*$ *form a subgroup of index* 2 *in* $\mathbf{F}_q^*$; *this subgroup is the kernel of the homomorphism* $x \mapsto x^{(q-1)/2}$ *with values in* $\{\pm 1\}$.

(In other terms, one has an exact sequence:

$$1 \to \mathbf{F}_q^{*2} \to \mathbf{F}_q^* \to \{\pm 1\} \to 1.)$$

Case (a) follows from the fact that $x \mapsto x^2$ is an automorphism of $\mathbf{F}_q$.

In case (b), let Ω be an algebraic closure of $\mathbf{F}_q$; if $x \in \mathbf{F}_q^*$, let $y \in \Omega$ be such that $y^2 = x$. We have:

$$y^{q-1} = x^{(q-1)/2} = \pm 1 \text{ since } x^{q-1} = 1.$$

For x to be a square in $\mathbf{F}_q$ it is necessary and sufficient that y belongs to $\mathbf{F}_q^*$, i.e. $y^{q-1} = 1$. Hence $\mathbf{F}_q^{*2}$ is the kernel of $x \mapsto x^{(q-1)/2}$. Moreover, since $\mathbf{F}_q^*$ is cyclic of order $q-1$, the index of $\mathbf{F}_q^{*2}$ is equal to 2.

3.2. *Legendre symbol (elementary case)*

Definition.—*Let* p *be a prime number* $\neq 2$, *and let* $x \in \mathbf{F}_p^*$. *The Legendre symbol of* x, *denoted by* $\left(\frac{x}{p}\right)$, *is the integer* $x^{(p-1)/2} = \pm 1$.

It is convenient to extend $\left(\frac{x}{p}\right)$ to all of $\mathbf{F}_p$ by putting $\left(\frac{0}{p}\right) = 0$. Moreover, if $x \in \mathbf{Z}$ has for image $x' \in \mathbf{F}_p$, one writes $\left(\frac{x}{p}\right) = \left(\frac{x'}{p}\right)$.

We have $\left(\frac{x}{p}\right)\left(\frac{y}{p}\right) = \left(\frac{xy}{p}\right)$: The Legendre symbol is a "character" (cf. chap. VI, §1). As seen in theorem 4, $\left(\frac{x}{p}\right) = 1$ is equivalent to $x \in \mathbf{F}_p^{*2}$; if $x \in \mathbf{F}_p^*$ has y as a square root in an algebraic closure of $\mathbf{F}_p$, then $\left(\frac{x}{p}\right) = y^{p-1}$.

Computation of $\left(\frac{x}{p}\right)$ *for* $x = 1, -1, 2$:

If n is an *odd* integer, let $\varepsilon(n)$ and $\omega(n)$ be the elements of $\mathbf{Z}/2\mathbf{Z}$ defined by:

$$\varepsilon(n) \equiv \frac{n-1}{2} \pmod 2 = \begin{cases} 0 \text{ if } n \equiv 1 \pmod 4 \\ 1 \text{ if } n \equiv -1 \pmod 4 \end{cases}$$

$$\omega(n) \equiv \frac{n^2-1}{8} \pmod 2 = \begin{cases} 0 \text{ if } n \equiv \pm 1 \pmod 8 \\ 1 \text{ if } n \equiv \pm 5 \pmod 8 \end{cases}$$

[The function ε is a homomorphism of the multiplicative group $(\mathbf{Z}/4\mathbf{Z})^*$ onto $\mathbf{Z}/2\mathbf{Z}$; similarly, ω is a homomorphism of $(\mathbf{Z}/8\mathbf{Z})^*$ onto $\mathbf{Z}/2\mathbf{Z}$.]

Theorem 5.—*The following formulas hold*:

i) $\left(\frac{1}{p}\right) = 1$

ii) $\left(\frac{-1}{p}\right) = (-1)^{\varepsilon(p)}$

iii) $\left(\frac{2}{p}\right) = (-1)^{\omega(p)}$.

Only the last deserves a proof. If α denotes a primitive 8th root of unity in an algebraic closure Ω of $\mathbf{F}_p$, the element $y = \alpha + \alpha^{-1}$ verifies $y^2 = 2$ (from $\alpha^4 = -1$ it follows that $\alpha^2 + \alpha^{-2} = 0$). We have

$$y^p = \alpha^p + \alpha^{-p}.$$

If $p \equiv \pm 1 \pmod 8$, this implies $y^p = y$, thus $\left(\frac{2}{p}\right) = y^{p-1} = 1$. If $p \equiv \pm 5 \pmod 8$, one finds $y^p = \alpha^5 + \alpha^{-5} = -(\alpha + \alpha^{-1}) = -y$. (This again follows from $\alpha^4 = -1$.) We deduce from this that $y^{p-1} = -1$, whence iii) follows.

Remark. Theorem 5 can be expressed in the following way:

-1 is a square (mod p) if and only if $p \equiv 1 \pmod 4$.
2 is a square (mod p) if and only if $p \equiv \pm 1 \pmod 8$.

3.3 *Quadratic reciprocity law*

Let l and p be two distinct prime numbers different from 2.

Theorem 6 (Gauss).— $\left(\frac{l}{p}\right) = \left(\frac{p}{l}\right)(-1)^{\varepsilon(l)\varepsilon(p)}$.

Let Ω be an algebraic closure of $\mathbf{F}_p$, and let $w \in \Omega$ be a primitive l-th root of unity. If $x \in \mathbf{F}_l$, the element w^x is well defined since $w^l = 1$. Thus we are able to form the "Gauss sum":

$$y = \sum_{x \in \mathbf{F}_l} \left(\frac{x}{l}\right) w^x.$$

Lemma 1.—$y^2 = (-1)^{\varepsilon(l)} l$.
(By abuse of notation l denotes also the image of l in the field $\mathbf{F}_p$.)

We have

$$y^2 = \sum_{x,z}\left(\frac{xz}{l}\right)w^{x+z} = \sum_{u\in\mathbf{F}_l} w^u\left(\sum_{t\in\mathbf{F}_l}\left(\frac{t(u-t)}{l}\right)\right).$$

Now if $t \neq 0$:

$$\left(\frac{t(u-t)}{l}\right) = \left(\frac{-t^2}{l}\right)\left(\frac{1-ut^{-1}}{l}\right) = (-1)^{\varepsilon(l)}\left(\frac{1-ut^{-1}}{l}\right),$$

and

$$(-1)^{\varepsilon(l)}y^2 = \sum_{u\in\mathbf{F}_l} C_u w^u,$$

where

$$C_u = \sum_{t\in\mathbf{F}_l^*}\left(\frac{1-ut^{-1}}{l}\right).$$

If $u = 0$, $C_0 = \sum_{t\in\mathbf{F}_l^*}\left(\frac{1}{l}\right) = l-1$; otherwise $s = 1-ut^{-1}$ runs over $\mathbf{F}_l - \{1\}$, and we have

$$C_u = \sum_{s\in\mathbf{F}_l}\left(\frac{s}{l}\right) - \left(\frac{1}{l}\right) = -\left(\frac{1}{l}\right) = -1,$$

since in $\mathbf{F}_l^*$ there are as many squares as non squares. Hence $\sum_{u\in\mathbf{F}_l} C_u w^u = l-1-\sum_{u\in\mathbf{F}_l^*} w^u = l$, which proves the lemma.

Lemma 2.—$y^{p-1} = \left(\frac{p}{l}\right)$

Since Ω is of characteristic p, we have

$$y^p = \sum_{x\in\mathbf{F}_l}\left(\frac{x}{p}\right)w^{xp} = \sum_{z\in\mathbf{F}_l}\left(\frac{zp^{-1}}{l}\right)w^z = \left(\frac{p^{-1}}{l}\right)y = \left(\frac{p}{l}\right)y;$$

hence $y^{p-1} = \left(\frac{p}{l}\right)$.

Theorem 6 is now immediate. Indeed, by lemmas 1 and 2,

$$\left(\frac{(-1)^{\varepsilon(l)}l}{p}\right) = y^{p-1} = \left(\frac{p}{l}\right)$$

and the second part of th. 5 proves that

$$\left(\frac{(-1)^{\varepsilon(l)}}{p}\right) = (-1)^{\varepsilon(l)\varepsilon(p)}.$$

Translation.—Write lRp if l is a square (mod p) (that is to say, if l is a "*quadratic residue*" modulo p) and lNp otherwise. Theorem 6 means that

$$lRp \Leftrightarrow pRl \quad \text{if } p \text{ or } l \equiv 1 \pmod 4$$

$$lRp \Leftrightarrow pNl \quad \text{if } p \text{ and } l \equiv -1 \pmod 4.$$

Remark. Theorem 6 can be used to compute Legendre symbols by successive reductions. Thus:

$$\left(\frac{29}{43}\right) = \left(\frac{43}{29}\right) = \left(\frac{14}{29}\right) = \left(\frac{2}{29}\right)\left(\frac{7}{29}\right) = -\left(\frac{7}{29}\right) = -\left(\frac{29}{7}\right) = -\left(\frac{1}{7}\right) = -1.$$

Appendix

Another proof of the quadratic reciprocity law (G. EISENSTEIN, *J. Crelle, 29*, 1845, pp. 177–184.)

i) *Gauss Lemma*

Let p be a prime number $\neq 2$, and let S be a subset of $\mathbf{F}_p^*$ such that $\mathbf{F}_p^*$ is the disjoint union of S and $-S$. In the following we take $S = \left\{1, \ldots, \frac{p-1}{2}\right\}$.

If $s \in S$ and $a \in \mathbf{F}_p^*$, we write as in the form $as = e_s(a)s_a$ with $e_s(a) = \pm 1$ and $s_a \in S$.

Lemma (Gauss).—$\left(\frac{a}{p}\right) = \prod_{s \in S} e_s(a)$.

Remark first that, if s and s' are two distinct elements of S, then $s_a \neq s'_a$ (for otherwise $s = \pm s'$ contrary to the choice of S). This shows that $s \mapsto s_a$ is a bijection of S onto itself. Multiplying the equalities $as = e_s(a)s_a$, we obtain

$$a^{(p-1)/2} \prod_{s \in S} s = \left(\prod_{s \in S} e_s(a)\right) \prod_{s \in S} s_a = \left(\prod_{s \in S} e_s(a)\right) \prod_{s \in S} s,$$

hence

$$a^{(p-1)/2} = \prod_{s \in S} e_s(a);$$

this proves the lemma since $\left(\frac{a}{p}\right) = a^{(p-1)/2}$ in $\mathbf{F}_p$.

Example.—Take $a = 2$ and $S = \left\{1, \ldots, \frac{p-1}{2}\right\}$. We have $e_s(2) = 1$ if $2s \leqq \frac{p-1}{2}$ and $e_s(2) = -1$ otherwise. From this we get $\left(\frac{2}{p}\right) = (-1)^{n(p)}$ where $n(p)$ is the number of integers s such that $\frac{p-1}{4} < s \leqq \frac{p-1}{2}$. If p is of the form $4k + 1$ (resp. $4k - 1$), then $n(p) = k$. Thus we recover the fact that $\left(\frac{2}{p}\right) = 1$ if $p \equiv \pm 1 \pmod 8$ and $\left(\frac{2}{p}\right) = -1$ if $p \equiv \pm 5 \pmod 8$, cf. th. 5.

ii) *A trigonometric lemma*

Lemma.—*Let m be a positive odd integer. One has*

$$\frac{\sin mx}{\sin x} = (-4)^{(m-1)/2} \prod_{1 \leq j \leq (m-1)/2} \left(\sin^2 x - \sin^2 \frac{2\pi j}{m} \right).$$

This is elementary (for instance, prove first that $\sin(mx)/\sin(x)$ is a polynomial of degree $(m-1)/2$ in $\sin^2 x$, then remark that this polynomial has for roots the $\sin^2 \frac{2\pi j}{m}$ with $1 \leq j \leq (m-1)/2$; the factor $(-4)^{(m-1)/2}$ is obtained by comparing coefficients of $e^{i(m-1)x}$ on both sides).

iii) *Proof of the quadratic reciprocity law*

Let l and p be two distinct prime numbers different from 2. Let

$$S = \{1, \ldots, (p-1)/2\}$$

as above. From Gauss' lemma, we get

$$\left(\frac{l}{p}\right) = \prod_{s \in S} e_s(l).$$

Now the equality $ls = e_s(l)s_l$ shows that

$$\sin \frac{2\pi}{p} ls = e_s(l) \sin \frac{2\pi}{p} s_l.$$

Multiplying these equalities, and taking into account that $s \mapsto s_l$ is a bijection, we get:

$$\left(\frac{l}{p}\right) = \prod_{s \in S} e_s(l) = \prod_{s \in S} \sin \frac{2\pi ls}{p} \Big/ \sin \frac{2\pi s}{p}.$$

By applying the trigonometric lemma with $m = l$, we can rewrite this:

$$\left(\frac{l}{p}\right) = \prod_{s \in S} (-4)^{(l-1)/2} \prod_{t \in T} \left(\sin^2 \frac{2\pi s}{p} - \sin^2 \frac{2\pi t}{l} \right)$$

$$= (-4)^{(l-1)(p-1)/4} \prod_{s \in S,\, t \in T} \left(\sin^2 \frac{2\pi s}{p} - \sin^2 \frac{2\pi t}{l} \right),$$

where T denotes the set of integers between 1 and $(l-1)/2$. Permuting the roles of l and p, we obtain similarly:

$$\left(\frac{p}{l}\right) = (-4)^{(l-1)(p-1)/4} \prod_{s \in S,\, t \in T} \left(\sin^2 \frac{2\pi t}{l} - \sin^2 \frac{2\pi s}{p} \right).$$

The factors giving $\left(\frac{l}{p}\right)$ and $\left(\frac{p}{l}\right)$ are identical up to sign. Since there are $(p-1)(l-1)/4$ of these, we find:

$$\left(\frac{l}{p}\right) = \left(\frac{p}{l}\right) (-1)^{(p-1)(l-1)/4}.$$

This is the quadratic reciprocity law, cf. th. 6.

Chapter II

p-Adic Fields

In this chapter p denotes a prime number.

§1. *The ring $\mathbf{Z}_p$ and the field $\mathbf{Q}_p$*

1.1. *Definitions*

For every $n \geqq 1$, let $A_n = \mathbf{Z}/p^n\mathbf{Z}$; it is the ring of classes of integers (mod p^n). An element of A_n defines in an obvious way an element of A_{n-1}; we thus obtain a homomorphism

$$\phi_n: A_n \to A_{n-1},$$

which is surjective and whose kernel is $p^{n-1}A_n$.

The sequence

$$\ldots \to A_n \to A_{n-1} \to \ldots \to A_2 \to A_1$$

forms a "projective system" indexed by the integers $\geqq 1$.

Definition 1.—*The ring of p-adic integers $\mathbf{Z}_p$ is the projective limit of the system (A_n, ϕ_n) defined above.*

By definition, an element of $\mathbf{Z}_p = \varprojlim (A_n, \phi_n)$ is a sequence $x = (\ldots, x_n, \ldots, x_1)$ with $x_n \in A_n$ and $\phi_n(x_n) = x_{n-1}$ if $n \geqq 2$. Addition and multiplication in $\mathbf{Z}_p$ are defined "coordinate by coordinate". In other words, $\mathbf{Z}_p$ is a *subring* of the product $\prod_{n \geqq 1} A_n$. If we give A_n the discrete topology and $\prod A_n$ the product topology, the ring $\mathbf{Z}_p$ inherits a topology which turns it into a *compact* space (since it is closed in a product of compact spaces).

1.2. *Properties of $\mathbf{Z}_p$*

Let $\varepsilon_n: \mathbf{Z}_p \to A_n$ be the function which associates to a p-adic integer x its n-th component x_n.

Proposition 1.—*The sequence $0 \to \mathbf{Z}_p \xrightarrow{p^n} \mathbf{Z}_p \xrightarrow{\varepsilon_n} A_n \to 0$ is an exact sequence of abelian groups.*

(Thus we can identify $\mathbf{Z}_p/p^n\mathbf{Z}_p$ with $A_n = \mathbf{Z}/p^n\mathbf{Z}$.)

Multiplication by p (hence also by p^n) is injective in $\mathbf{Z}_p$; indeed, if $x = (x_n)$ is a p-adic integer such that $px = 0$, we have $px_{n+1} = 0$ for all n, and x_{n+1} is of the form $p^n y_{n+1}$ with $y_{n+1} \in A_{n+1}$; since $x_n = \phi_{n+1}(x_{n+1})$, we see that x_n is also divisible by p^n, hence, is zero.

It is clear that the kernel of ε_n contains $p^n\mathbf{Z}_p$; conversely, if $x = (x_m)$ belongs to $\ker(\varepsilon_m)$, one has $x_m \equiv 0 \pmod{p^n}$ for all $m \geqq n$ which means that there exists a well defined element y_{m-n} of A_{m-n} such that its image under the isomorphism $A_{m-n} \to p^n\mathbf{Z}/p^m\mathbf{Z} \subset A_m$ satisfies $x_m = p^n y_{m-n}$. The y_i define an element y of $\mathbf{Z}_p = \varprojlim A_i$, and one checks immediately that $p^n y = x$, which proves the proposition.

Proposition 2.—(a) *For an element of $\mathbf{Z}_p$ (resp. of A_n) to be invertible it is necessary and sufficient that it is not divisible by p.*

(b) *If $\mathbf{U}$ denotes the group of invertible elements of $\mathbf{Z}_p$, every nonzero element of $\mathbf{Z}_p$ can be written uniquely in the form $p^n u$ with $u \in \mathbf{U}$ and $n \geqq 0$.* (An element of $\mathbf{U}$ is called a *p-adic unit.*)

It suffices to prove (a) for A_n; the case of $\mathbf{Z}_p$ will follow. Now, if $x \in A_n$ does not belong to pA_n, its image in $A_1 = \mathbf{F}_p$ is not zero, thus invertible: hence there exists $y, z \in A_n$ such that $xy = 1 - pz$, hence

$$xy(1+pz+\ldots+p^{n-1}z^{n-1}) = 1,$$

which proves that x is invertible.

On the other hand, if $x \in \mathbf{Z}_p$ is not zero, there exists a largest integer n such that $x_n = \varepsilon_n(x)$ is zero; then $x = p^n u$ with u not divisible by p, hence $u \in \mathbf{U}$ by (a). The uniqueness of the decomposition is clear.

Notation.—Let x be a nonzero element of $\mathbf{Z}_p$; write x in the form $p^n u$ with $u \in \mathbf{U}$. The integer n is called the *p-adic valuation* of x and denoted by $v_p(x)$. We put $v_p(0) = +\infty$ and we have

$$v_p(xy) = v_p(x)+v_p(y), \quad v_p(x+y) \geqq \inf(v_p(x), v_p(y))$$

It follows easily from these formulas that $\mathbf{Z}_p$ is an *integral domain.*

Proposition 3.—*The topology on $\mathbf{Z}_p$ can be defined by the distance*

$$d(x, y) = e^{-v_p(x-y)}.$$

The ring $\mathbf{Z}_p$ is a complete metric space in which $\mathbf{Z}$ is dense.

The ideals $p^n\mathbf{Z}_p$ form a basis of neighborhoods of 0; since $x \in p^n\mathbf{Z}_p$ is equivalent to $v_p(x) \geqq n$, the topology on $\mathbf{Z}_p$ is defined by the distance $d(x, y) = e^{-v_p(x-y)}$. Since $\mathbf{Z}_p$ is compact, it is complete. Finally, if $x = (x_n)$ is an element of $\mathbf{Z}_p$, and if $y_n \in \mathbf{Z}$ is such that $y_n \equiv x_n \pmod{p^n}$, then $\lim . y_n = x$, which proves that $\mathbf{Z}$ is dense in $\mathbf{Z}_p$.

1.3. *The field* $\mathbf{Q}_p$

Definition 2.—*The field of p-adic numbers, denoted by $\mathbf{Q}_p$, is the field of fractions of the ring $\mathbf{Z}_p$.*

One sees immediately that $\mathbf{Q}_p = \mathbf{Z}_p[p^{-1}]$. Every element x of $\mathbf{Q}_p^*$ can be written uniquely in the form $p^n u$ with $n \in \mathbf{Z}$, $u \in \mathbf{U}$; here again, n is called the p-adic valuation of x and is denoted by $v_p(x)$. One has $v_p(x) \geqq 0$ if and only if $x \in \mathbf{Z}_p$.

Proposition 4.—*The field* $\mathbf{Q}_p$, *with the topology defined by* $d(x, y) = e^{-v_p(x-y)}$, *is locally compact, and contains* $\mathbf{Z}_p$ *as an open subring; the field* $\mathbf{Q}$ *is dense in* $\mathbf{Q}_p$.

This is clear.

Remarks.—1) We could have defined $\mathbf{Q}_p$ (resp. $\mathbf{Z}_p$) as the *completion* of $\mathbf{Q}$ (resp. $\mathbf{Z}$) for the p-adic distance d.

2) The distance d satisfies the "*ultrametric*" inequality

$$d(x, z) \leqq \sup (d(x, y), d(y, z)).$$

From this one sees that a sequence u_n has a limit if and only if

$$\lim . (u_{n+1} - u_n) = 0;$$

similarly, a series converges if and only if its general term tends to 0.

§2. *p-adic equations*

2.1. *Solutions*

Lemma.—*Let* $\ldots \to D_n \to D_{n-1} \to \ldots \to D_1$ *be a projective system, and let* $D = \varprojlim . D_n$ *be its projective limit. If the* D_n *are finite and nonempty, then* D *is nonempty.*

The fact that $D \neq \varnothing$ is clear if the $D_n \to D_{n-1}$ are surjective; we are going to reduce the lemma to this special case. For this, denote by $D_{n,p}$ the image of D_{n+p} in D_n; for fixed n, the $D_{n,p}$ form a decreasing family of finite nonempty subsets; hence this family is *stationary*, i.e. $D_{n,p}$ is independent of p for p large enough. Let E_n be this limit value of the $D_{n,p}$. One checks immediately that $D_n \to D_{n-1}$ carries E_n *onto* E_{n-1}; since the E_n are nonempty, we have $\varprojlim . E_n \neq \varnothing$ by the remark made at the beginning; hence, *a fortiori* $\varprojlim . D_n \neq \varnothing$.

Notation.—If $f \in \mathbf{Z}_p[X_1, \ldots, X_m]$ is a polynomial with coefficients in $\mathbf{Z}_p$, and if n is an integer $\geqq 1$, we denote by f_n the polynomial with coefficients in A_n deduced from f by reduction (mod p^n).

Proposition 5.—*Let* $f^{(i)} \in \mathbf{Z}_p[X_1, \ldots, X_m]$ *be polynomials with p-adic integer coefficients. The following are equivalent:*

i) *The* $f^{(i)}$ *have a common zero in* $(\mathbf{Z}_p)^m$.

ii) *For all* $n > 1$, *the polynomials* $f_n^{(i)}$ *have a common zero in* $(A_n)^m$.

Let D (resp. D_n) be the set of common zeros of the $f^{(i)}$ (resp. $f_n^{(i)}$). The D_n are finite and we have $D = \varprojlim . D_n$. By the above lemma, D is nonempty if and only if the D_n are nonempty; hence the proposition.

A point $x = (x_1, \ldots, x_m)$ of $(\mathbf{Z}_p)^m$ is called *primitive* if one of the x_i is invertible, that is, if the x_i are not all divisible by p. One defines in a similar way the primitive elements of $(A_n)^m$.

Proposition 6.—*Let $f^{(i)} \in \mathbf{Z}_p[X_1, \ldots, X_m]$ be homogeneous polynomials with p-adic integer coefficients. The following are equivalent:*

a) *The $f^{(i)}$ have a non trivial common zero in $(\mathbf{Q}_p)^m$.*
b) *The $f^{(i)}$ have a common primitive zero in $(\mathbf{Z}_p)^m$.*
c) *For all $n > 1$, the $f_n^{(i)}$ have a common primitive zero in $(A_n)^m$.*

The implication b) $\Rightarrow$ a) is trivial. Conversely, if $x = (x_1, \ldots, x_m)$ is a nontrivial common zero of the $f^{(i)}$, put

$$h = \inf(v_p(x_1), \ldots, v_p(x_m)) \quad \text{and} \quad y = p^{-h}x.$$

It is clear that y is a primitive element of $(\mathbf{Z}_p)^m$, and that it is a common zero of the $f^{(i)}$. Hence b) $\Leftrightarrow$ a).

The equivalence of b) and c) follows from the above lemma.

2.2. *Amelioration of approximate solutions.*

We are concerned with passing from a solution (mod p^n) to a true solution (i.e. with coefficients in $\mathbf{Z}_p$). One uses the following lemma (p-adic analogue of "Newton's method"):

Lemma.—*Let $f \in \mathbf{Z}_p[X]$ and let f' be its derivative. Let $x \in \mathbf{Z}_p$, $n, k \in \mathbf{Z}$ such that $0 \leqq 2k < n$, $f(x) \equiv 0 \pmod{p^n}$, $v_p(f'(x)) = k$. Then there exists $y \in \mathbf{Z}_p$ such that*

$$f(y) \equiv 0 \pmod{p^{n+1}}, \quad v_p(f'(y)) = k, \quad \textit{and} \quad y \equiv x \pmod{p^{n-k}}.$$

Take y of the form $x+p^{n-k}z$ with $z \in \mathbf{Z}_p$. By Taylor's formula we have

$$f(y) = f(x)+p^{n-k}zf'(x)+p^{2n-2k}a \quad \text{with } a \in \mathbf{Z}_p.$$

By hypothesis $f(x) = p^n b$ and $f'(x) = p^k c$ with $b \in \mathbf{Z}_p$ and $c \in \mathbf{U}$. This allows us to choose z in such a way that

$$b+zc \equiv 0 \pmod{p}.$$

From this we get

$$f(y) = p^n(b+zc)+p^{2n-2k}a \equiv 0 \pmod{p^{n+1}}$$

since $2n-2k > n$. Finally Taylor's formula applied to f' shows that $f'(y) \equiv p^k c \pmod{p^{n-k}}$; since $n-k > k$, we see that $v_p(f'(y)) = k$.

Theorem 1.—*Let $f \in \mathbf{Z}_p[X_1, \ldots, X_m]$, $x = (x_i) \in (\mathbf{Z}_p)^m$, $n, k \in \mathbf{Z}$ and j an integer such that $0 \leq j \leq m$. Suppose that $0 < 2k < n$ and that*

$$f(x) \equiv 0 \pmod{p^n} \quad \textit{and} \quad v_p\left(\frac{\partial f}{\partial X_j}(x)\right) = k.$$

Then there exists a zero y of f in $(\mathbf{Z}_p)^m$ which is congruent to x modulo p^{n-k}.

Suppose first that $m = 1$. By applying the above lemma to $x^{(0)} = x$, we obtain $x^{(1)} \in \mathbf{Z}_p$ congruent to $x^{(0)} \pmod{p^{n-k}}$ and such that

$$f(x^{(1)}) \equiv 0 \pmod{p^{n+1}} \text{ and } v_p(f'(x^{(1)})) = k.$$

We can apply the lemma to $x^{(1)}$, after replacing n by $n+1$. Arguing inductively, we construct in this way a sequence $x^{(0)}, \ldots, x^{(q)}, \ldots$ such that

$$x^{(q+1)} \equiv x^{(q)} \pmod{p^{n+q-k}}, \quad f(x^{(q)}) \equiv 0 \pmod{p^{n+q}}.$$

This is a Cauchy sequence. If y is its limit, we have $f(y) = 0$ and $y \equiv x \pmod{p^{n-k}}$, hence the theorem for $m = 1$.

The case $m > 1$ reduces to the case $m = 1$ by modifying only x_j. More precisely, let $\tilde{f} \in \mathbf{Z}_p[X_j]$ be the polynomial in one variable obtained by replacing X_i, $i \neq j$, by x_i. What has just been proven can be applied to $\tilde{f}$ and x_j; this shows the existence of $y_j \equiv x_j \pmod{p^{n-k}}$ such that $\tilde{f}(y_j) = 0$. If one puts $y_i = x_i$ for $i \neq j$, the element $y = (y_i)$ satisfies the desired condition.

Corollary 1.—*Every simple zero of the reduction modulo p of a polynomial f lifts to a zero of f with coefficients in* $\mathbf{Z}_p$.

(If g is a polynomial over a field k, a zero x of g is called *simple* if at least one of the partial derivatives $\partial g/\partial X_j$ is nonzero at x.)

This is the special case $n = 1$, $k = 0$.

Corollary 2.—*Suppose* $p \neq 2$. *Let* $f(X) = \Sigma a_{ij} X_i X_j$ *with* $a_{ij} = a_{ji}$ *be a quadratic form with coefficients in* $\mathbf{Z}_p$ *whose discriminant* $\det(a_{ij})$ *is invertible. Let* $a \in \mathbf{Z}_p$. *Every primitive solution of the equation* $f(x) \equiv a \pmod{p}$ *lifts to a true solution.*

In view of cor. 1, it suffices to show that x does not annihilate all the partial derivatives of f modulo p. Now $\dfrac{\partial f}{\partial X_i} = 2\Sigma_j a_{ij} X_j$; since $\det(a_{ij}) \not\equiv 0 \pmod{p}$ and x is primitive, one of these partial derivatives is $\not\equiv 0 \pmod{p}$.

Corollary 3.—*Suppose* $p = 2$. *Let* $f = \Sigma a_{ij} X_i X_j$ *with* $a_{ij} = a_{ji}$ *be a quadratic form with coefficients in* $\mathbf{Z}_2$ *and let* $a \in \mathbf{Z}_2$. *Let x be a primitive solution of* $f(x) \equiv a \pmod{8}$. *We can lift x to a true solution provided x does not annihilate all the* $\dfrac{\partial f}{\partial X_j}$ *modulo 4; this last condition is fulfilled if* $\det(a_{ij})$ *is invertible.*

The first assertion follows from the theorem applied to $n = 3$, $k = 1$; the second can be proved as in the case $p \neq 2$ (taking into account the factor 2).

§3. *The multiplicative group of* $\mathbf{Q}_p$

3.1. *The filtration of the group of units*

Let $\mathbf{U} = \mathbf{Z}_p^*$ be the group of p-adic units. For every $n \geqq 1$, put $\mathbf{U}_n = 1 + p^n \mathbf{Z}_p$; this is the kernel of the homomorphism $\varepsilon_n \colon \mathbf{U} \to (\mathbf{Z}/p^n\mathbf{Z})^*$. In particular, the quotient $\mathbf{U}/\mathbf{U}_1$ can be identified with $\mathbf{F}_p^*$, hence is cyclic of order $p-1$ (cf. Chap. I, th. 2). The $\mathbf{U}_n$ form a decreasing sequence of open subgroups of $\mathbf{U}$, and $\mathbf{U} = \varprojlim . \mathbf{U}/\mathbf{U}_n$. If $n \geqq 1$, the map

$$(1 + p^n x) \mapsto (x \text{ modulo } p)$$

defines an isomorphism $\mathbf{U}_n/\mathbf{U}_{n+1} \to \mathbf{Z}/p\mathbf{Z}$; this follows from the formula:

$$(1+p^n x)(1+p^n y) \equiv 1+p^n(x+y) \pmod{p^{n+1}}.$$

We see from this, by induction on n, that $\mathbf{U}_1/\mathbf{U}_n$ has order p^{n-1}.

Lemma.—*Let $0 \to A \to E \to B \to 0$ be an exact sequence of commutative groups (denoted additively) with A and B finite with orders a and b prime to each other. Let B' be the set of $x \in E$ such that $bx = 0$. The group E is the direct sum of A and B'. Moreover B' is the only subgroup of E isomorphic to B.*

Since a and b are relatively prime, there exist $r, s \in \mathbf{Z}$ such that $ar+bs = 1$. If $x \in A \cap B'$, then $ax = bx = 0$, hence $(ar+bs)\,x = x = 0$; and $A \cap B' = 0$. Moreover, all $x \in E$ can be written $x = arx+bsx$; since $bB' = 0$, we have $bE \subset A$, hence $bsx \in A$; on the other hand, from $abE = 0$ follows that $arx \in B'$. Hence we see that $E = A \oplus B'$ and the projection $E \to B$ defines an isomorphism of B' onto B. Conversely, if B'' is a subgroup of E isomorphic to B, we have $bB'' = 0$ hence $B'' \subset B'$ and $B'' = B'$ because these groups have the same order.

Proposition 7.—*One has $\mathbf{U} = \mathbf{V} \times \mathbf{U}_1$ where $\mathbf{V} = \{x \in \mathbf{U} | x^{p-1} = 1\}$ is the unique subgroup of $\mathbf{U}$ isomorphic to $\mathbf{F}_p^*$.*

One applies the lemma to the exact sequences

$$1 \to \mathbf{U}_1/\mathbf{U}_n \to \mathbf{U}/\mathbf{U}_n \to \mathbf{F}_p^* \to 1,$$

which is allowable because the order of $\mathbf{U}_1/\mathbf{U}_n$ is p^{n-1} and the order of $\mathbf{F}_p^*$ is $p-1$. From this, one concludes that $\mathbf{U}/\mathbf{U}_n$ contains a unique subgroup $\mathbf{V}_n$ isomorphic to $\mathbf{F}_p^*$ and the projection

$$\mathbf{U}/\mathbf{U}_n \to \mathbf{U}/\mathbf{U}_{n-1}$$

carries $\mathbf{V}_n$ isomorphically onto $\mathbf{V}_{n-1}$. Since $\mathbf{U} = \varprojlim . \mathbf{U}/\mathbf{U}_n$, we get from this, by passage to the limit, a subgroup $\mathbf{V}$ of $\mathbf{U}$ isomorphic to $\mathbf{F}_p^*$. One has $\mathbf{U} = \mathbf{V} \times \mathbf{U}_1$; the uniqueness of $\mathbf{V}$ follows from that of the $\mathbf{V}_n$.

Corollary.—*The field $\mathbf{Q}_p$ contains the $(p-1)$th roots of unity.*

Remarks—1) The group $\mathbf{V}$ is called the group of *multiplicative representatives* of the elements of $\mathbf{F}_p^*$.

2) The existence of $\mathbf{V}$ can also be proved by applying cor. 1 of th. 1 to the equation $x^{q-1}-1 = 0$.

3.2. *Structure of the group* $\mathbf{U}_1$

Lemma.—*Let $x \in \mathbf{U}_n - \mathbf{U}_{n+1}$ with $n \geqq 1$ if $p \neq 2$ and $n \geqq 2$ if $p = 2$. Then $x^p \in \mathbf{U}_{n+1} - \mathbf{U}_{n+2}$.*

By hypothesis, one has $x = 1+kp^n$ with $k \not\equiv 0 \pmod p$. The binomial formula gives

$$x^p = 1+kp^{n+1}+ \dots +k^p p^{np}$$

and the exponents in the terms not written are $\geqq 2n+1$, hence also $\geqq n+2$.

Moreover $np \geqq n+2$ (due to the fact that $n \geqq 2$ if $p = 2$). This shows that

$$x^p \equiv 1+kp^{n+1} \pmod{p^{n+2}}$$

hence $x^p \in \mathbf{U}_{n+1} - \mathbf{U}_{n+2}$.

Proposition 8.—*If $p \neq 2$, $\mathbf{U}_1$ is isomorphic to $\mathbf{Z}_p$. If $p = 2$, $\mathbf{U}_1 = \{\pm 1\} \times \mathbf{U}_2$ and $\mathbf{U}_2$ is isomorphic to $\mathbf{Z}_2$.*

Consider first the case $p \neq 2$. Choose an element $\alpha \in \mathbf{U}_1 - \mathbf{U}_2$, for example $\alpha = 1+p$. By the above lemma, we have $\alpha^{p^i} \in \mathbf{U}_{i+1} - \mathbf{U}_{i+2}$. Let α_n be the image of α in $\mathbf{U}_1/\mathbf{U}_n$; we have $(\alpha_n)^{p^{n-2}} \neq 1$ and $(\alpha_n)^{p^{n-1}} = 1$. But $\mathbf{U}_1/\mathbf{U}_n$ is of order p^{n-1}; hence it is a *cyclic* group, generated by α_n. Now, denote by $\theta_{n,\alpha}$ the isomorphism $z \mapsto \alpha_n^z$ of $\mathbf{Z}/p^{n-1}\mathbf{Z}$ onto $\mathbf{U}_1/\mathbf{U}_n$. The diagram

$$\begin{array}{ccc} \mathbf{Z}/p^n\mathbf{Z} & \xrightarrow{\theta_{n+1,\alpha}} & \mathbf{U}_1/\mathbf{U}_{n+1} \\ \downarrow & & \downarrow \\ \mathbf{Z}/p^{n-1}\mathbf{Z} & \xrightarrow{\theta_{n,\alpha}} & \mathbf{U}_1/\mathbf{U}_n \end{array}$$

is commutative. From this one sees that the $\theta_{n,\alpha}$ define an isomorphism θ of $\mathbf{Z}_p = \varprojlim \mathbf{Z}/p^{n-1}\mathbf{Z}$ onto $\mathbf{U}_1 = \varprojlim \mathbf{U}_1/\mathbf{U}_n$, hence the proposition for $p \neq 2$.

Suppose now that $p = 2$. Choose $\alpha \in \mathbf{U}_2 - \mathbf{U}_3$, that is $\alpha \equiv 5 \pmod 8$. Define as above isomorphisms

$$\theta_{n,\alpha}\colon \mathbf{Z}/2^{n-2}\mathbf{Z} \to \mathbf{U}_2/\mathbf{U}_n,$$

hence an isomorphism $\theta_\alpha\colon \mathbf{Z}_2 \to \mathbf{U}_2$. On the other hand, the homomorphism

$$\mathbf{U}_1 \to \mathbf{U}_1/\mathbf{U}_2 \simeq \mathbf{Z}/2\mathbf{Z}$$

induces an isomorphism of $\{\pm 1\}$ onto $\mathbf{Z}/2\mathbf{Z}$. From this we get

$$\mathbf{U}_1 = \{\pm 1\} \times \mathbf{U}_2, \quad \text{q.e.d.}$$

Theorem 2.—*The group $\mathbf{Q}_p^*$ is isomorphic to $\mathbf{Z} \times \mathbf{Z}_p \times \mathbf{Z}/(p-1)\mathbf{Z}$ if $p \neq 2$ and to $\mathbf{Z} \times \mathbf{Z}_2 \times \mathbf{Z}/2\mathbf{Z}$ if $p = 2$.*

Every element $x \in \mathbf{Q}_p^*$ can be written uniquely in the form $x = p^n u$ with $n \in \mathbf{Z}$ and $u \in \mathbf{U}$. Hence $\mathbf{Q}_p^* \simeq \mathbf{Z} \times \mathbf{U}$. Moreover, prop. 7 proves that $\mathbf{U} = \mathbf{V} \times \mathbf{U}_1$ where $\mathbf{V}$ is cyclic of order $p-1$, and the structure of $\mathbf{U}_1$ is given by prop. 8.

3.3. *Squares in $\mathbf{Q}_p^*$*

Theorem 3.—*Suppose $p \neq 2$ and let $x = p^n u$ be an element of $\mathbf{Q}_p^*$, with $n \in \mathbf{Z}$ and $u \in \mathbf{U}$. For x to be a square it is necessary and sufficient that n is even and the image $\bar{u}$ of u in $\mathbf{F}_p^* = \mathbf{U}/\mathbf{U}_1$ is a square.*

(The last condition means that the *Legendre symbol* $\left(\frac{\bar{u}}{p}\right)$ of $\bar{u}$ is equal to 1. We write in the following $\left(\frac{u}{p}\right)$ instead of $\left(\frac{\bar{u}}{p}\right)$.)

Decompose u in the form $u = v.u_1$ with $v \in \mathbf{V}$ and $u_1 \in \mathbf{U}_1$. The decomposition $\mathbf{Q}_p^* \simeq \mathbf{Z} \times \mathbf{V} \times \mathbf{U}_1$ of th. 2 proves that x is a square if and only if n is even and v and u_1 are squares; but $\mathbf{U}_1$ is isomorphic to $\mathbf{Z}_p$ and 2 is invertible in $\mathbf{Z}_p$; all the elements of $\mathbf{U}_1$ are then squares. Since $\mathbf{V}$ is isomorphic to $\mathbf{F}_p^*$, the theorem follows.

Corollary.—*If $p \neq 2$, the group $\mathbf{Q}_p^*/\mathbf{Q}_p^{*2}$ is a group of type* (2, 2). *It has for representatives* $\{1, p, u, up\}$ *where $u \in \mathbf{U}$ is such that* $\left(\frac{u}{p}\right) = -1$.

This is clear.

Theorem 4.—*For an element $x = p^n u$ of $\mathbf{Q}_2^*$ to be a square it is necessary and sufficient that n is even and $u \equiv 1 \pmod 8$.*

The decomposition $\mathbf{U} = \{\pm 1\} \times \mathbf{U}_2$ shows that u is a square if and only if u belongs to $\mathbf{U}_2$ and is a square in $\mathbf{U}_2$. Now the isomorphism $\theta \colon \mathbf{Z}_2 \to \mathbf{U}_2$ constructed in the proof of prop. 8 carries $2^n\mathbf{Z}_2$ onto $\mathbf{U}_{n+2}$. Taking $n = 1$, we see that the set of squares of $\mathbf{U}_2$ is equal to $\mathbf{U}_3$. An element $u \in \mathbf{U}$ is then a square if and only if it is congruent to 1 modulo 8, hence the theorem.

Remark.—The fact that every element of $\mathbf{U}_3$ is a square follows also from cor. 3 of th. 1 applied to the quadratic form X^2.

Corollary.—*The group $\mathbf{Q}_2^*/\mathbf{Q}_2^{*2}$ is of type* (2, 2, 2). *It has for representatives* $\{\pm 1, \pm 5, \pm 2, \pm 10\}$.

This follows from the fact that $\{\pm 1, \pm 5\}$ is a system of representatives for $\mathbf{U}/\mathbf{U}_3$.

Remarks.

1) For $p = 2$, define homomorphisms $\varepsilon, \omega \colon \mathbf{U}/\mathbf{U}_3 \to \mathbf{Z}/2\mathbf{Z}$ by means of the formulas of chap. I, n° 3.2:

$$\varepsilon(z) \equiv \frac{z-1}{2} \pmod 2 = \begin{cases} 0 & \text{if} \quad z \equiv 1 \pmod 4 \\ 1 & \text{if} \quad z \equiv -1 \pmod 4 \end{cases}$$

$$\omega(z) \equiv \frac{z^2-1}{8} \pmod 2 = \begin{cases} 0 & \text{if} \quad z \equiv \pm 1 \pmod 8 \\ 1 & \text{if} \quad z \equiv \pm 5 \pmod 8. \end{cases}$$

The map ε defines an isomorphism of $\mathbf{U}/\mathbf{U}_2$ onto $\mathbf{Z}/2\mathbf{Z}$ and the map ω an isomorphism of $\mathbf{U}_2/\mathbf{U}_3$ onto $\mathbf{Z}/2\mathbf{Z}$. The pair (ε, ω) defines thus an *isomorphism* of $\mathbf{U}/\mathbf{U}_3$ onto $\mathbf{Z}/2\mathbf{Z} \times \mathbf{Z}/2\mathbf{Z}$; in particular a 2-adic unit z is a square if and only if $\varepsilon(z) = \omega(z) = 0$.

2) Theorems 3 and 4 show that $\mathbf{Q}_p^{*2}$ is an open subgroup of $\mathbf{Q}_p^*$.

Chapter III

Hilbert Symbol

§1. *Local properties*

In this paragraph, k denotes either the field $\mathbf{R}$ of real numbers or the field $\mathbf{Q}_p$ of p-adic numbers (p being a prime number).

1.1. *Definition and first properties*

Let $a, b \in k^*$. We put:

$(a, b) = 1$ if $z^2 - ax^2 - by^2 = 0$ has a solution $(z, x, y) \neq (0, 0, 0)$ in k^3.

$(a, b) = -1$ otherwise.

The number $(a, b) = \pm 1$ is called the *Hilbert symbol* of a and b relative to k. It is clear that (a, b) does not change when a and b are multiplied by squares; thus the Hilbert symbol defines a *map* from $k^*/k^{*2} \times k^*/k^{*2}$ into $\{\pm 1\}$.

Proposition 1.—*Let $a, b \in k^*$ and let $k_b = k(\sqrt{b})$. For $(a, b) = 1$ it is necessary and sufficient that a belongs to the group Nk_b^* of norms of elements of k_b^*.*

If b is the square of an element c, the equation $z^2 - ax^2 - by^2 = 0$ has $(c, 0, 1)$ for a solution, hence $(a, b) = 1$, and the proposition is clear in this case since $k_b = k$ and $Nk_b^* = k^*$. Otherwise, k_b is quadratic over k; if β denotes a square root of b, every element $\xi \in k_b$ can be written $z + \beta y$ with $y, z \in k$ and the norm $N(\xi)$ of ξ is equal to $z^2 - by^2$. If $a \in Nk_b^*$, there exist $y, z \in k$ such that $a = z^2 - by^2$, so that the quadratic form $z^2 - ax^2 - by^2$ has a zero $(z, 1, y)$ and we have $(a, b) = 1$.

Conversely, if $(a, b) = 1$, this form has a zero $(z, x, y) \neq (0, 0, 0)$. One has $x \neq 0$, for otherwise b would be a square. From this we see that a is the norm of $\dfrac{z}{x} + \beta \dfrac{y}{x}$.

Proposition 2.—*The Hilbert symbol satisfies the formulas:*

i) $(a, b) = (b, a)$ *and* $(a, c^2) = 1$,

ii) $(a, -a) = 1$ *and* $(a, 1-a) = 1$,

iii) $(a, b) = 1 \Rightarrow (aa', b) = (a', b)$,

iv) $(a, b) = (a, -ab) = (a, (1-a)b)$.

(In these formulas a, a', b, c denote elements of k^*; one supposes $a \neq 1$ when the formula contains the term $1-a$.)

Formula i) is obvious. If $b = -a$ (resp. if $b = 1-a$) the quadratic form $z^2 - ax^2 - by^2$ has for zero $(0, 1, 1)$ (resp. $(1, 1, 1)$); thus $(a, b) = 1$, which proves ii). If $(a, b) = 1$, a is contained in the subgroup Nk_b^*, cf. prop. 1;

we then have $a' \in Nk_b^* \Leftrightarrow aa' \in Nk_b^*$, which proves iii). Formula iv) follows from i), ii), iii).

Remark.—Formula iii) is a particular case of

$$\text{v)} \qquad (aa', b) = (a, b)(a', b),$$

which expresses the *bilinearity* of the Hilbert symbol; this formula will be proved in the following section.

1.2. *Computation of* (a, b)

Theorem 1.—*If* $k = \mathbf{R}$, *we have* $(a, b) = 1$ *if* a *or* b *is* >0, *and* $(a, b) = -1$ *if* a *and* b *are* <0.

If $k = \mathbf{Q}_p$ *and if we write* a, b *in the form* $p^\alpha u, p^\beta v$ *where* u *and* v *belong to the group* $\mathbf{U}$ *of* p*-adic units, we have*

$$(a, b) = (-1)^{\alpha\beta\varepsilon(p)}\left(\frac{u}{p}\right)^\beta\left(\frac{v}{p}\right)^\alpha \qquad \text{if } p \neq 2$$

$$(a, b) = (-1)^{\varepsilon(u)\varepsilon(v) + \alpha\omega(v) + \beta\omega(u)} \qquad \text{if } p = 2.$$

[Recall that $\left(\frac{u}{p}\right)$ denotes the *Legendre symbol* $\left(\frac{\bar{u}}{p}\right)$ where $\bar{u}$ is the image of u by the homomorphism of reduction modulo p: $\mathbf{U} \to \mathbf{F}_p^*$. As for $\varepsilon(u)$ and $\omega(u)$, they denote respectively the class modulo 2 of $\frac{u-1}{2}$ and of $\frac{u^2-1}{8}$, cf. Chap. II, n° 3.3.]

Theorem 2.—*The Hilbert symbol is a nondegenerate bilinear form on the* $\mathbf{F}_2$*-vector space* k^*/k^{*2}.

[The bilinearity of (a, b) is just formula v) mentioned at the end of n° 1.1. The assertion "(a, b) is nondegenerate" means that, if $b \in k^*$ is such that $(a, b) = 1$ for all $a \in k^*$, one has $b \in k^{*2}$.]

Corollary.—*If* b *is not a square, the group* Nk_b^* *defined in prop.* 1 *is a subgroup of index* 2 *in* k^*.

The homomorphism ϕ_b: $k^* \to \{\pm 1\}$ defined by $\phi_b(a) = (a, b)$ has kernel Nk_b^* by prop. 1; moreover, ϕ_b is surjective since (a, b) is nondegenerate. Hence ϕ_b defines an isomorphism of k^*/Nk_b^* onto $\{\pm 1\}$; the corollary follows from this.

Remark.—More generally, let L be a finite extension of k which is galoisian and whose Galois group G is *commutative*. One can prove that k^*/NL^* *is isomorphic to* G and that the knowledge of the group NL^* *determines* L. These are two of the main results of the so-called "local class field theory."

Proof of theorems 1 *and* 2.

The case $k = \mathbf{R}$ is trivial. Note that k^*/k^{*2} is then a vector space of dimension 1 (over the field $\mathbf{F}_2$) having $\{1, -1\}$ for representatives.

Suppose now that $k = \mathbf{Q}_p$.

Lemma.—*Let $v \in \mathbf{U}$ be a p-adic unit. If the equation $z^2 - px^2 - vy^2 = 0$ has a nontrivial solution in $\mathbf{Q}_p$, it has a solution (z, x, y) such that $z, y \in \mathbf{U}$ and $x \in \mathbf{Z}_p$.*

By prop. 6 of chap. II, n° 2.1, the given equation has a *primitive* solution (z, x, y). Let us show that this solution has the desired property. If it did not, we would have either $y \equiv 0 \pmod p$ or $z \equiv 0 \pmod p$; since $z^2 - vy^2 \equiv 0 \pmod p$ and $v \not\equiv 0 \pmod p$, we would have both $y \equiv 0 \pmod p$ and $z \equiv 0 \pmod p$, hence $px^2 \equiv 0 \pmod{p^2}$, i.e. $x \equiv 0 \pmod p$ contrary to the primitive character of (z, x, y).

We now return to the proof of theorem 1, *and we suppose first that $p \neq 2$.*

It is clear that the exponents α and β come in only by their residue modulo 2; in view of the symmetry of the Hilbert symbol, there are only three cases to consider:

1) $\alpha = 0, \beta = 0$. We must check that $(u, v) = 1$. Now the equation

$$z^2 - ux^2 - vy^2 = 0$$

has a nontrivial solution modulo p (chap. I, §2, cor. 2 to th. 3); since the discriminant of this quadratic form is a p-adic unit, the above solution lifts to a p-adic solution (chap. II, n° 2.2, cor. 2 to th. 1); hence $(u, v) = 1$.

2) $\alpha = 1, \beta = 0$. We must check that $(pu, v) = \left(\frac{v}{p}\right)$. Since $(u, v) = 1$ we have $(pu, v) = (p, v)$ by formula iii) of prop. 2; thus it suffices to check that $(p, v) = \left(\frac{v}{p}\right)$. This is clear if v is a square, the two terms being equal to 1. Otherwise $\left(\frac{v}{p}\right) = -1$, see chap. II, n° 3.3, th. 3. Then the above lemma shows that $z^2 - px^2 - vy^2$ does not have a nontrivial zero and so $(p, v) = -1$.

3) $\alpha = 1, \beta = 1$. We must check that $(pu, pv) = (-1)^{(p-1)/2}\left(\frac{u}{p}\right)\left(\frac{v}{p}\right)$.

Formula iv) of prop. 2 shows that:

$$(pu, pv) = (pu, -p^2uv) = (pu, -uv).$$

By what we have just seen, $(pu, pv) = \left(\frac{-uv}{p}\right)$, from which the desired result follows since $\left(\frac{-1}{p}\right) = (-1)^{(p-1)/2}$.

Once theorem 1 is established (for $p \neq 2$), theorem 2 follows from it, since the formula giving (a, b) is bilinear; in order to prove the nondegeneracy, it suffices to exhibit, for all $a \in k^*/k^{*2}$ distinct from the neutral element, an element b such that $(a, b) = -1$. By cor. to th. 3 of chap. II, n° 3.3, we can take $a = p$, u or up with $u \in \mathbf{U}$ such that $\left(\frac{u}{p}\right) = -1$; then we choose for b respectively, u, p, and u.

The case $p = 2$. Here again, α and β come in only by their residue modulo 2 and there are three cases to consider:

1) $\alpha = 0$, $\beta = 0$. We must check that $(u, v) = 1$ if u or v is congruent to 1 (mod 4) and $(u, v) = -1$ otherwise. Suppose first that $u \equiv 1 \pmod 4$. Then $u \equiv 1 \pmod 8$ or $u \equiv 5 \pmod 8$. In the first case u is a square (chap. II, n° 3.3, th. 4) and we do have $(u, v) = 1$. In the second case we have $u+4v \equiv 1 \pmod 8$ and there exists $w \in \mathbf{U}$ such that $w^2 = u+4v$; the form $z^2-ux^2-vy^2$ has thus $(w, 1, 2)$ for a zero and we do have $(u, v) = 1$. Let us now suppose $u \equiv v \equiv -1 \pmod 4$; if (z, x, y) is a primitive solution of $z^2-ux^2-vy^2 = 0$, then $z^2+x^2+y^2 \equiv 0 \pmod 4$; but the squares of $\mathbf{Z}/4\mathbf{Z}$ are 0 and 1; this congruence implies that x, y, z are congruent to 0 (mod 2), which contradicts the hypothesis of primitivity. Thus we have $(u, v) = -1$ in this case.

2) $\alpha = 1$, $\beta = 0$. We must check that $(2u, v) = (-1)^{\varepsilon(u)\varepsilon(v)+\omega(v)}$. First let us show that $(2, v) = (-1)^{\omega(v)}$, i.e. that $(2, v) = 1$ is equivalent to $v \equiv \pm 1 \pmod 8$. By the above lemma if $(2, v) = 1$, there exists $x, y, z \in \mathbf{Z}_2$ such that $z^2-2x^2-vy^2 = 0$ and $y, z \not\equiv 0 \pmod 2$. Then we have $y^2 = z^2 \equiv 1 \pmod 8$, hence $1-2x^2-v \equiv 0 \pmod 8$. But the only squares modulo 8 are 0, 1, and 4; from this we get $v \equiv \pm 1 \pmod 8$. Conversely, if $v \equiv 1 \pmod 8$, v is a square and $(2, v) = 1$; if $v \equiv -1 \pmod 8$, the equation $z^2-2x^2-vy^2 = 0$ has $(1, 1, 1)$ for a solution modulo 8, and this approximate solution lifts to a true solution (chap. II, n° 2.2, cor. 3 to th. 1); thus we have $(2, v) = 1$.

We show next that $(2u, v) = (2, v)\,(u, v)$; by prop. 2, this is true if $(2, v) = 1$ or $(u, v) = 1$. The remaining case is $(2, v) = (u, v) = -1$, i.e. $v \equiv 3 \pmod 8$ and $u \equiv 3$ or $-1 \pmod 8$; after multiplying u and v by squares, we can suppose that $u = -1$, $v = 3$ or $u = 3$, $v = -5$; now the equations

$$z^2+2x^2-3y^2 = 0 \text{ and } z^2-6x^2+5y^2 = 0$$

have for solution $(1, 1, 1)$; thus we have $(2u, v) = 1$.

3) $\alpha = 1$, $\beta = 1$. We must check that

$$(2u, 2v) = (-1)^{\varepsilon(u)\varepsilon(v)+\omega(u)+\omega(v)}.$$

Now formula iv) of prop. 2 shows that

$$(2u, 2v) = (2u, -4uv) = (2u, -uv).$$

By what we have just seen, we have

$$(2u, 2v) = (-1)^{\varepsilon(u)\varepsilon(-uv)+\omega(-uv)}.$$

Since $\varepsilon(-1) = 1$, $\omega(-1) = 0$ and $\varepsilon(u)\,(1+\varepsilon(u)) = 0$, the above exponent is equal to $\varepsilon(u)\varepsilon(v)+\omega(u)+\omega(v)$, which proves theorem 1. The bilinearity of (a, b) follows from the formula giving this symbol, since ε and ω are homomorphisms. The nondegeneracy is checked on the representatives $\{u, 2u\}$ with $u = 1, 5, -1$ or -5. Indeed, we have $(5, 2u) = -1$ and

$$(-1, -1) = (-1, -5) = -1.$$

Remark.—Write (a, b) in the form $(-1)^{[a,b]}$ with $[a, b] \in \mathbf{Z}/2\mathbf{Z}$. Then $[a, b]$ is a symmetric bilinear form on k^*/k^{*2} with values in $\mathbf{Z}/2\mathbf{Z}$ and th. 1 gives its matrix with respect to some basis of k^*/k^{*2}:
— For $k = \mathbf{R}$, it is the matrix (1).
— For $k = \mathbf{Q}_p$, $p \neq 2$, with basis $\{p, u\}$ where $\left(\frac{u}{p}\right) = -1$, it is the matrix $\begin{pmatrix} 0 & 1 \\ 1 & 0 \end{pmatrix}$ if $p \equiv 1 \pmod 4$ and $\begin{pmatrix} 1 & 1 \\ 1 & 0 \end{pmatrix}$ if $p \equiv 3 \pmod 4$.
— For $k = \mathbf{Q}_2$, with basis $\{2, -1, 5\}$ it is the matrix $\begin{pmatrix} 0 & 0 & 1 \\ 0 & 1 & 0 \\ 1 & 0 & 0 \end{pmatrix}$.

§2. *Global properties*

The field $\mathbf{Q}$ of rational numbers embeds as a subfield into each of the fields $\mathbf{Q}_p$ and $\mathbf{R}$. If $a, b \in \mathbf{Q}^*$, $(a, b)_p$ (resp. $(a, b)_\infty$) denotes the Hilbert symbol of their images in $\mathbf{Q}_p$ (resp. $\mathbf{R}$). We define V to be the set of prime numbers together with the symbol ∞, and make the convention that $\mathbf{Q}_\infty = \mathbf{R}$, hence $\mathbf{Q}$ is dense in $\mathbf{Q}_v$ for all $v \in V$.

2.1. *Product formula*

Theorem 3 (Hilbert).—*If $a, b \in \mathbf{Q}^*$, we have $(a, b)_v = 1$ for almost all $v \in V$ and*

$$\prod_{v \in V} (a, b)_v = 1.$$

(The expression "almost all $v \in V$" means "all the elements of V except a finite number".)

Since the Hilbert symbols are bilinear, it suffices to prove the theorem when a or b are equal to -1 or to a prime number. In each case, theorem 1 gives the value of $(a, b)_v$. We find

1) $a = -1$, $b = -1$. One has $(-1, -1)_\infty = (-1, -1)_2 = -1$ and $(-1, -1)_p = 1$ if $p \neq 2, \infty$; the product is equal to 1.

2) $a = -1$, $b = l$ with l prime. If $l = 2$, one has $(-1, 2)_v = 1$ for all $v \in V$; if $l \neq 2$, one has $(-1, l)_v = 1$ if $v \neq 2, l$ and $(-1, l)_2 = (-1, l)_l = (-1)^{\varepsilon(l)}$. The product is equal to 1.

3) $a = l$, $b = l'$ with l, l' primes. If $l = l'$, formula iv) of prop. 2 shows that $(l, l)_v = (-1, l)_v$ for all $v \in V$ and we are reduced to the case considered above. If $l \neq l'$ and if $l' = 2$, one has $(l, 2)_v = 1$ for $v \neq 2, l$ and

$$(l, 2)_2 = (-1)^{\omega(l)}, \; (l, 2)_l = \left(\frac{2}{l}\right) = (-1)^{\omega(l)}, \text{ cf. chap. I, n° 3.2, th. 5.}$$

If l and l' are distinct and different from 2, one has $(l, l')_v = 1$ for $v \neq 2, l, l'$ and

$$(l, l')_2 = (-1)^{\varepsilon(l)\varepsilon(l')}, \; (l, l')_l = \left(\frac{l'}{l}\right), \; (l, l')_{l'} = \left(\frac{l}{l'}\right);$$

but by the quadratic reciprocity law (chap. I, n° 3.3, th. 6) one has

$$\left(\frac{l'}{l}\right)\left(\frac{l}{l'}\right) = (-1)^{\varepsilon(l)\varepsilon(l')};$$

hence the product is equal to 1. This completes the proof.

Remark.—The product formula is essentially equivalent to the quadratic reciprocity law. Its interest comes mainly from the fact that it extends to *all algebraic number fields* (the set V being replaced by the set of "places" of the field).

2.2. *Existence of rational numbers with given Hilbert symbols*

Theorem 4.—*Let $(a_i)_{i\in I}$ be a finite family of elements in $\mathbf{Q}^*$ and let $(\varepsilon_{i,v})_{i\in I, v\in V}$ be a family of numbers equal to ± 1. In order that there exists $x \in \mathbf{Q}^*$ such that $(a_i, x)_v = \varepsilon_{i,v}$ for all $i \in I$ and all $v \in V$, it is necessary and sufficient that the following conditions be satisfied:*
(1) *Almost all the $\varepsilon_{i,v}$ are equal to 1.*
(2) *For all $i \in I$ we have $\prod_{v\in V} \varepsilon_{i,v} = 1$.*
(3) *For all $v \in V$ there exists $x_v \in \mathbf{Q}_v^*$ such that $(a_i, x_v)_v = \varepsilon_{i,v}$ for all $i \in I$.*

The necessity of (1) and (2) follows from theorem 3; that of (3) is trivial (take $x_v = x$).

To prove the sufficiency of these conditions, we need the following three lemmas:

Lemma 1 ("Chinese remainder theorem").—*Let $a_1, \ldots, a_n, m_1, \ldots, m_n$ be integers with the m_i being pairwise relatively prime. There exists an integer a such that $a \equiv a_i \pmod{m_i}$ for all i.*

Let m be the product of m_i. Bezout theorem shows that the canonical homomorphism

$$\mathbf{Z}/m\mathbf{Z} \to \prod_{i=1}^{i=n} \mathbf{Z}/m_i\mathbf{Z}$$

is an isomorphism. The lemma follows from this.

Lemma 2 ("Approximation theorem").—*Let S be a finite subset of V. The image of $\mathbf{Q}$ in $\prod_{v\in S} \mathbf{Q}_v$ is dense in this product (for the product topology of those of $\mathbf{Q}_v$).*

Being free to enlarge S, we can suppose that $S = \{\infty, p_1, \ldots p_n\}$ where the p_i are distinct prime numbers and we must prove that $\mathbf{Q}$ is dense in $\mathbf{R}\times\mathbf{Q}_{p_1}\times \ldots \times\mathbf{Q}_{p_n}$. Let $(x_\infty, x_1, \ldots, x_n)$ be a point of this product and let us show that this point is adherent to $\mathbf{Q}$. After multiplying by some integer, we may suppose that $x_i \in \mathbf{Z}_{p_i}$ for $1 \leqq i \leqq n$. Now one has to prove that, for all $\varepsilon > 0$ and all integers $N > 0$, there exists $x \in \mathbf{Q}$ such that

$$|x - x_\infty| \leqq \varepsilon \quad \text{and} \quad v_{p_i}(x - x_i) \geqq N \quad \text{for } i = 1, \ldots, n.$$

By lemma 1 applied to $m_i = p_i^N$, there exists $x_0 \in \mathbf{Z}$ such that $v_{p_i}(x_0 - x_i) \geqq N$ for all i. Choose now an integer $q \geqq 2$ which is prime to all the p_i (for example a prime number). The rational numbers of the form a/q^m, $a \in \mathbf{Z}$, $m \geqq 0$, are *dense* in $\mathbf{R}$ (this follows simply from the fact $q^m \to \infty$ when $m \to \infty$). Choose such a number $u = a/q^m$ with

$$|x_0 - x_\infty + up_1^N \dots p^N| \leqq \varepsilon.$$

The rational number $x = x_0 + up_1^N \dots p_n^N$ has the desired property.

Lemma 3 ("Dirichlet theorem").—*If a and m are relatively prime integers* $\geqq 1$, *there exist infinitely many primes p such that* $p \equiv a \pmod{m}$.

The proof will be given in chap. VI; the reader can check that it uses none of the results of chapters III, IV, and V.

Now come back to theorem 4, and let $(\varepsilon_{i,v})$ be a family of numbers equal to ± 1 and satisfying conditions (1), (2), and (3). After multiplying the a_i by the square of some integer, we can suppose that all the a_i are *integers*. Let S be the subset of V made of ∞, 2, and the prime factors of a_i; let T be the set of $v \in V$ such that there exists $i \in I$ with $\varepsilon_{i,v} = -1$; these two sets are finite. We distinguish two cases:

1) *We have* $S \cap T = \varnothing$.

Put

$$a = \prod_{\substack{l \in T \\ l \neq \infty}} l \quad \text{and} \quad m = 8 \prod_{\substack{l \in S \\ l \neq 2, \infty}} l.$$

Because $S \cap T = \varnothing$, the integers a and m are relatively prime and, by lemma 3, there exists a prime number $p \equiv a \pmod{m}$ with $p \notin S \cup T$. We are going to show that $x = ap$ has the desired property, i.e. $(a_i, x)_v = \varepsilon_{i,v}$ for all $i \in I$ and $v \in V$.

If $v \in S$, we have $\varepsilon_{i,v} = 1$ since $S \cap T = \varnothing$, and we must check that $(a_i, x)_v = 1$. If $v = \infty$, this follows from $x > 0$; if v is a prime number l, we have $x \equiv a^2 \pmod{m}$, hence $x \equiv a^2 \pmod{8}$ for $l = 2$ and $x \equiv a^2 \pmod{l}$ for $l \neq 2$; since x and a are l-adic units, this shows that x is a square in $\mathbf{Q}_l^*$ (cf. chap. II, n° 3.3) and we have $(a_i, x)_v = 1$.

If $v = l$ is not in S, a_i is an l-adic unit. Since $l \neq 2$ we have

$$(a_i, b)_l = \left(\frac{a_i}{l}\right)^{v_l(b)} \quad \text{for all } b \in \mathbf{Q}_l^*, \text{ cf. th. 1.}$$

If $l \notin T \cup \{p\}$, x is an l-adic unit, hence $v_l(x) = 0$ and the above formula shows that $(a_i, x)_l = 1$; on the other hand, we have $\varepsilon_{i,l} = 1$ because $l \notin T$. If $l \in T$, we have $v_l(x) = 1$; moreover, condition (3) shows that there exists $x_l \in \mathbf{Q}_l^*$ such that $(a_i, x_l)_l = \varepsilon_{i,l}$ for all $i \in I$; since one of the $\varepsilon_{i,l}$ is equal to -1 (because l belongs to T), we have $v_l(x_l) \equiv 1 \pmod{2}$ hence

$$(a_i, x)_l = \left(\frac{a_i}{l}\right) = (a_i, x_l)_l = \varepsilon_{i,l} \qquad \text{for all } i \in I.$$

There remains the case $l = p$, which we deduce from the others using the product formula:

$$(a_i, x)_p = \prod_{v \neq p} (a_i, x)_v = \prod_{v \neq p} \varepsilon_{i,v} = \varepsilon_{i,p}.$$

This completes the proof of theorem 4 in the case $S \cap T = \varnothing$.

2) *General case.*

We know that the squares of $\mathbf{Q}_v^*$ form an open subgroup of $\mathbf{Q}_v^*$, cf. chap. II, n° 3.3. By lemma 2, there exists $x' \in \mathbf{Q}^*$ such that x'/x_v is a square in $\mathbf{Q}_v^*$ for all $v \in S$. In particular $(a_i, x')_v = (a_i, x_v)_v = \varepsilon_{i,v}$ for all $v \in S$. If we set $\eta_{i,v} = \varepsilon_{i,v}(a_i, x')_v$, the family $(\eta_{i,v})$ verifies conditions (1), (2), (3) and moreover $\eta_{i,v} = 1$ if $v \in S$. By 1) above there exists $y \in \mathbf{Q}^*$ such that $(a_i, y)_v = \eta_{i,v}$ for all $i \in I$ and all $v \in V$. If we set $x = yx'$, it is clear that x has the desired properties.

Chapter IV

Quadratic Forms over $\mathbf{Q}_p$ and over $\mathbf{Q}$

§1. *Quadratic forms*

1.1. *Definitions*

First recall the general notion of a *quadratic form* (see BOURBAKI, *Alg.*, chap. IX, 3, n° 4).

Definition 1.—*Let V be a module over a commutative ring A. A function $Q: V \to A$ is called a quadratic form on V if:*

1) $Q(ax) = a^2 Q(x)$ *for* $a \in A$ *and* $x \in V$
2) *The function* $(x, y) \mapsto Q(x+y) - Q(x) - Q(y)$ *is a bilinear form.*

Such a pair (V, Q) is called a *quadratic module.* In this chapter, we limit ourselves to the case where the ring A is *a field k of characteristic $\neq 2$*; the A-module V is then a k-vector space; we suppose that its dimension is *finite.*

We put:

$$x.y = \tfrac{1}{2}\{Q(x+y) - Q(x) - Q(y)\};$$

this makes sense since the characteristic of k is different from 2. The map $(x, y) \mapsto x.y$ is a *symmetric bilinear* form on V, called the *scalar product* associated with Q. One has $Q(x) = x.x$. This establishes a bijective correspondence between *quadratic forms* and *symmetric bilinear forms* (it would not be so in characteristic 2).

If (V, Q) and (V', Q') are two quadratic modules, a linear map $f: V \to V'$ such that $Q' \circ f = Q$ is called a *morphism* (or *metric morphism*) of (V, Q) into (V', Q'); then $f(x).f(y) = x.y$ for all $x, y \in V$.

Matrix of a quadratic form.—Let $(e_i)_{1 \leq i \leq n}$ be a basis of V. The matrix of Q with respect to this basis is the matrix $A = (a_{ij})$ where $a_{ij} = e_i.e_j$; it is symmetric. If $x = \Sigma x_i e_i$ is an element of V, then

$$Q(x) = \sum_{i,j} a_{ij} x_i x_j,$$

which shows that $Q(x)$ is a "quadratic form" in $x_1, \ldots, x_n$ in the usual sense.

If we change the basis (e_i) by means of an invertible matrix X, the matrix A' of Q with respect to the new basis is $X.A.{}^tX$ where ${}^tX$ denotes the transpose of X. In particular

$$\det(A') = \det(A).\det(X)^2,$$

which shows that $\det(A)$ is *determined up to multiplication by an element of* k^{*2}; it is called the *discriminant* of Q and denoted by $\mathrm{disc}(Q)$.

1.2. *Orthogonality*

Let (V, Q) be a quadratic module over k. Two elements x, y of V are called *orthogonal* if $x.y = 0$. The set of elements orthogonal to a subset H of V is denoted by H^0; it is a vector subspace of V. If V_1 and V_2 are two vector subspaces of V, they are said to be *orthogonal* if $V_1 \subset V_2^0$, i.e. if $x \in V_1$, $y \in V_2$ implies $x.y = 0$.

The orthogonal complement V^0 of V itself is called the *radical* (or the kernel) of V and denoted by $\text{rad}(V)$. Its codimension is called the *rank* of Q. If $V^0 = 0$ we say that Q is *nondegenerate*; this is equivalent to saying that the discriminant of Q is $\neq 0$ (in which case we view it as an element of the group k^*/k^{*2}).

Let U be a vector subspace of V, and let U^* be the dual of U. Let q_U: $V \to U^*$ be the function which associates to each $x \in V$ the linear form $(y \in U \mapsto x.y)$. The kernel of q_U is U^0. In particular we see that Q is nondegenerate if and only if q_V: $V \to V^*$ is an *isomorphism*.

Definition 2.—*Let $U_1, \ldots, U_m$ be vector subspaces of V. One says that V is the orthogonal direct sum of the U_i if they are pairwise orthogonal and if V is the direct sum of them. One writes then*:

$$V = U_1 \hat{\oplus} \ldots \hat{\oplus} U_m.$$

Remark.—If $x \in V$ has for components x_i in U_i,

$$Q(x) = Q_1(x_1) + \ldots + Q_m(x_m),$$

where $Q_i = Q|U_i$ denotes the restriction of Q to U_i. Conversely if (U_i, Q_i) is a family of quadratic modules, the formula above endows $V = \oplus U_i$ with a quadratic form Q, called the *direct sum* of the Q_i, and one has $V = U_1 \hat{\oplus} \ldots \hat{\oplus} U_m$.

Proposition 1.—*If U is a supplementary subspace of* $\text{rad}(V)$ *in V, then* $V = U \hat{\oplus} \text{rad}(V)$.

This is clear.

Proposition 2.—*Suppose (V, Q) is nondegenerate. Then*:

i) *All metric morphisms of V into a quadratic module (V', Q') are injective.*

ii) *For all vector subspaces U of V, we have*

$$U^{00} = U, \ \dim U + \dim U^0 = \dim V, \ \text{rad}(U) = \text{rad}(U^0) = U \cap U^0.$$

The quadratic module U is nondegenerate if and only if U^0 is nondegenerate, in which case $V = U \hat{\oplus} U^0$.

iii) *If V is the orthogonal direct sum of two subspaces, they are nondegenerate and each of them is orthogonal to the other.*

If f: $V \to V'$ is a metric morphism, and if $f(x) = 0$, we have

$$x.y = f(x).f(y) = 0 \ \text{ for all } \ y \in V;$$

this implies $x = 0$ because (V, Q) is nondegenerate.

If U is a vector subspace of V, the homomorphism $q_U\colon V \to U^*$ defined above is *surjective*; indeed, it is obtained by composing $q_V\colon V \to V^*$ with the canonical surjection $V^* \to U^*$ and we have supposed that q_V is bijective. Thus we have an exact sequence:

$$0 \to U^0 \to V \to U^* \to 0,$$

hence $\dim V = \dim U^* + \dim U^0 = \dim U + \dim U^0$.

This shows that U and U^{00} have the same dimension; since U is contained in U^{00}, we have $U = U^{00}$; the formula $\operatorname{rad}(U) = U \cap U^0$ is clear; applying it to U^0, and taking into account that $U^{00} = U$, we get $\operatorname{rad}(U^0) = \operatorname{rad}(U)$ and at the same time the last assertion of ii). Finally iii) is trivial.

1.3. *Isotropic vectors*

Definition 3.—*An element x of a quadratic module (V, Q) is called isotropic if $Q(x) = 0$. A subspace U of V is called isotropic if all its elements are isotropic.*

We have evidently:

$$U \text{ isotropic} \Leftrightarrow U \subset U^0 \Leftrightarrow Q|U = 0.$$

Definition 4.—*A quadratic module having a basis formed of two isotropic elements x,y such that $x.y \neq 0$ is called a hyperbolic plane.*

After multiplying y by $1/x.y$, we can suppose that $x.y = 1$. Then the matrix of the quadratic form with respect to x, y is simply $\begin{pmatrix} 0 & 1 \\ 1 & 0 \end{pmatrix}$; its discriminant is -1 (in particular, it is nondegenerate).

Proposition 3.—*Let x be an isotropic element $\neq 0$ of a nondegenerate quadratic module (V, Q). Then there exists a subspace U of V which contains x and which is a hyperbolic plane.*

Since V is nondegenerate, there exists $z \in V$ such that $x.z = 1$. The element $y = 2z - (z.z)x$ is isotropic and $x.y = 2$. The subspace $U = kx + ky$ has the desired property.

Corollary.—*If (V, Q) is nondegenerate and contains a nonzero isotropic element, one has $Q(V) = k$.*

(In other words, for all $a \in k$, there exists $v \in V$ such that $Q(v) = a$.)

In view of the proposition, it suffices to give the proof when V is a hyperbolic plane with basis x, y with x, y isotropic and $x.y = 1$. If $a \in k$, then $a = Q\left(x + \frac{a}{2}y\right)$, and from this $Q(V) = k$.

1.4. *Orthogonal basis*

Definition 5.—*A basis $(e_1, \ldots, e_n)$ of a quadratic module (V, Q) is called orthogonal if its elements are pairwise orthogonal, i.e. if $V = ke_1 \hat{\oplus} \ldots \hat{\oplus} ke_n$.*

This amounts to saying that the matrix of Q with respect to this basis is a diagonal matrix:

$$\begin{pmatrix} a_1 & 0 & \dots & 0 \\ 0 & a_2 & \dots & 0 \\ & \dots & \dots & \\ 0 & 0 & \dots & a_n \end{pmatrix}.$$

If $x = \Sigma x_i e_i$, one has $Q(x) = a_1 x_1^2 + \dots + a_n x_n^2$.

Theorem 1.—*Every quadratic module (V, Q) has an orthogonal basis.*

We use induction on $n = \dim V$, the case $n = 0$ being trivial. If V is isotropic, all bases of V are orthogonal. Otherwise, choose an element $e_1 \in V$ such that $e_1.e_1 \neq 0$. The orthogonal complement H of e_1 is a hyperplane and since e_1 does not belong to H, one has $V = ke_1 \hat{\oplus} H$; in view of the inductive hypothesis, H has an orthogonal basis $(e_2, \dots, e_n)$, and $(e_1, \dots, e_n)$ has the desired property.

Definition 6.—*Two orthogonal bases*

$$\mathbf{e} = (e_1, \dots, e_n) \text{ and } \mathbf{e}' = (e'_1, \dots, e'_n)$$

of V are called contiguous if they have an element in common (i.e. *if there exist i and j such that $e_i = e'_j$*).

Theorem 2.—*Let (V, Q) be a nondegenerate quadratic module of dimension $\geqq 3$, and let $\mathbf{e} = (e_1, \dots e_n)$, $\mathbf{e}' = (e'_1, \dots e'_n)$ be two orthogonal bases of V. There exists a finite sequence $\mathbf{e}^{(0)}, \mathbf{e}^{(1)}, \dots, \mathbf{e}^{(m)}$ of orthogonal bases of V such that $\mathbf{e}^{(0)} = \mathbf{e}$, $\mathbf{e}^{(m)} = \mathbf{e}'$ and $\mathbf{e}^{(i)}$ is contiguous with $\mathbf{e}^{(i+1)}$ for $0 \leqq i < m$.*

(One says that $\mathbf{e}^{(0)}, \dots, \mathbf{e}^{(m)}$ is a *chain* of orthogonal bases contiguously relating $\mathbf{e}$ to $\mathbf{e}'$)

We distinguish three cases:

i) $(e_1.e_1)(e'_1.e'_1) - (e_1.e'_1)^2 \neq 0$

This amounts to saying that e_1 and e'_1 are not proportional and that the *plane $P = ke_1 + ke'_1$ is nondegenerate.* There exist then ε_2, ε'_2 such that

$$P = ke_1 \hat{\oplus} k\varepsilon_2 \text{ and } P = ke'_1 \hat{\oplus} k\varepsilon'_2.$$

Let H be the orthogonal complement of P; since P is nondegenerate, we have $V = H \hat{\oplus} P$, see prop. 2. Let $(e''_3, \dots, e''_n)$ be an orthogonal basis of H. One can then relate $\mathbf{e}$ to $\mathbf{e}'$ by means of the chain;

$$\mathbf{e} \to (e_1, \varepsilon_2, e''_3, \dots, e''_n) \to (e'_1, \varepsilon'_2, e''_3, \dots, e''_n) \to \mathbf{e}',$$

hence the theorem in this case.

ii) $(e_1.e_1)(e'_2.e'_2) - (e_1.e'_2)^2 \neq 0$

Same proof replacing e'_1 by e'_2.

iii) $(e_1.e_1)(e'_i.e'_i) - (e_1.e'_i)^2 = 0$ for $i = 1, 2$.

We prove first:

Lemma.—*There exists $x \in k$ such that $e_x = e'_1 + xe'_2$ is nonisotropic and generates with e_1 a nondegenerate plane.*

We have $e_x.e_x = e_1'.e_1' + x^2(e_2'.e_2')$; we must thus take x^2 distinct from $-(e_1'.e_1')/(e_2'.e_2')$. Moreover, for e_x to generate with e_1 a nondegenerate plane, it is necessary and sufficient that

$$(e_1.e_1)\,(e_x.e_x) - (e_1.e_x)^2 \neq 0.$$

If we make this explicit, taking into account the hypothesis iii), we find that the left hand side is $-2x(e_1.e_1')\,(e_1.e_2')$. Now hypothesis iii) implies $e_1.e_i' \neq 0$ for $i = 1, 2$. We see thus that e_x verifies the conditions of the lemma if and only if we have $x \neq 0$ and $x^2 \neq -(e_1'.e_1')/(e_2'.e_2')$. This eliminates at most three values of x; if k has at least 4 elements, we can find one such x. There remains the case $k = \mathbf{F}_3$ (the case $k = \mathbf{F}_2$ is excluded because $\mathrm{char}(k) \neq 2$). But, then, all non-zero squares are equal to 1 and hypothesis iii) can be written $(e_1.e_1)\,(e_i'.e_i') = 1$ for $1 = 1, 2$; the expression $(e_1'.e_1')/(e_2'.e_2')$ is thus equal to 1, and, in order to realize the condition $x^2 \neq 0, -1$, it suffices to take $x = 1$.

This being so, let us choose $e_x = e_1' + xe_2'$ verifying the conditions of the lemma. Since e_x is not isotropic, there exists e_2'' such that (e_x, e_2'') is an orthogonal basis of $ke_1' \hat{\oplus} ke_2'$. Let us put

$$e'' = (e_x, e_2'', e_3', \ldots, e_n');$$

it is an orthogonal basis of V. Since $ke_1 + ke_x$ is a nondegenerate plane, part i) of the proof shows that one can relate $\boldsymbol{e}$ to $\boldsymbol{e}''$ by a chain of contiguous bases; since $\boldsymbol{e}'$ and $\boldsymbol{e}''$ are contiguous, the theorem follows.

1.5. *Witt's theorem*

Let (V, Q) and (V', Q') be two *nondegenerate* quadratic modules; let U be a subvector space of V, and let

$$s\colon U \to V'$$

be an *injective* metric morphism of U into V'. We try to extend s to a subspace larger than U and if possible to all of V. We begin with the case where U is degenerate:

Lemma.—*If U is degenerate, we can extend s to an injective metric morphism $s_1\colon U_1 \to V'$ where U_1 contains U as a hyperplane.*

Let x be a non-zero element of $\mathrm{rad}(U)$, and l a linear form on U such that $l(x) = 1$. Since V is nondegenerate, there exists $y \in V$ such that $l(u) = u.y$ for all $u \in U$; we can moreover assume that $y.y = 0$ (replace y by $y - \lambda x$, with $\lambda = \frac{1}{2}y.y$). The space $U_1 = U \oplus ky$ contains U as a hyperplane.

The same construction, applied to $U' = s(U)$, $x' = s(x)$ and $l' = l \circ s^{-1}$ gives $y' \in V'$ and $U_1' = U' \oplus ky'$. One then checks that the linear map $s_1\colon U_1 \to V'$ which coincides with s on U and carries y onto y' is a metric isomorphism of U_1 onto U_1'.

Theorem 3 (Witt).—*If (V, Q) and (V', Q') are isomorphic and nondegenerate, every injective metric morphism*

$$s\colon U \to V'$$

of a subvector space U of V can be extended to a metric isomorphism of V onto V'.

Since V and V' are isomorphic, we can suppose that $V = V'$. Moreover, by applying the above lemma, we are reduced to the case where U is *nondegenerate*. We argue then by induction on dim U.

If dim $U = 1$, U is generated by a nonisotropic element x; if $y = s(x)$, we have $y.y = x.x$. One can choose $\varepsilon = \pm 1$ such that $x+\varepsilon y$ is not isotropic; otherwise, we would have

$$2x.x+2x.y = 2x.x-2x.y = 0$$

which would imply $x.x = 0$. Choose such an ε, and let H be the orthogonal complement of $z = x+\varepsilon y$; we have $V = kz \oplus H$. Let σ be the "symmetry with respect to H", i.e. the automorphism of V which is the identity on H and which changes z to $-z$. Since $x-\varepsilon y$ is contained in H, we have

$$\sigma(x-\varepsilon y) = x-\varepsilon y \text{ and } \sigma(x+\varepsilon y) = -x-\varepsilon y,$$

hence $\sigma(x) = -\varepsilon y$, and the automorphism $-\varepsilon\sigma$ extends s.

If dim $U > 1$, we decompose U in the form $U_1 \hat{\oplus} U_2$, with $U_1, U_2 \neq 0$. By the inductive hypothesis, the restriction s_1 of s to U_1 extends to an automorphism σ_1 of V; after replacing s by $\sigma_1^{-1} \circ s$, one can thus suppose that s is the identity on U_1. Then the morphism s carries U_2 into the orthogonal complement V_1 of U_1; by the inductive hypothesis, the restriction of s to U_2 extends to an automorphism σ_2 of V_1; the automorphism σ of V which is the identity on U_1 and σ_2 on V_1 has the desired property.

Corollary.—*Two isomorphic subspaces of a nondegenerate quadratic module have isomorphic orthogonal complements.*

One extends an isomorphism between the two subspaces to an automorphism of the module and restricts it to the orthogonal complements.

1.6. *Translations*

Let $f(X) = \sum_{i=1}^{n} a_{ii}X_i^2 + 2\sum_{i<j} a_{ij}X_iX_j$ be a quadratic form in n variables over k; we put $a_{ij} = a_{ji}$ if $i > j$ so that the matrix $A = (a_{ij})$ is symmetric. The pair (k^n, f) is a quadratic module, *associated to* f (or to the matrix A).

Definition 7.—*Two quadratic forms f and f' are called equivalent if the corresponding modules are isomorphic.*

Then we write $f \sim f'$. If A and A' are the matrices of f and f', this amounts to saying that there exists an invertible matrix X such that $A' = X.A.{}^tX$, see n° 1.1.

Let $f(X_1, \ldots, X_n)$ and $g(X_1, \ldots, X_m)$ be two quadratic forms; we will denote $f \dotplus g$ (or simply $f+g$ if no confusion is possible) the quadratic form

$$f(X_1, \ldots, X_n)+g(X_{n+1}, \ldots, X_{n+m})$$

in $n+m$ variables. This operation corresponds to that of *orthogonal sum* (see def. 2, n° 1.2). We write similarly $f \dot{-} g$ (or simply $f-g$) for $f+(-g)$. Here are some examples of translations:

Definition 4'—*A form $f(X_1, X_2)$ in two variables is called hyperbolic if we have*

$$f \sim X_1 X_2 \sim X_1^2 - X_2^2.$$

(This means that the module (k^2, f) corresponding is a *hyperbolic plane*, cf. def. 4).

We say that a form $f(X_1, \ldots, X_n)$ represents an element a of k if there exists $x \in k^n$, $x \neq 0$, such that $f(x) = a$. In particular f represents 0 if and only if the corresponding quadratic module contains a non-zero isotropic element.

Proposition 3'.—*If f represents 0 and is nondegenerate, one has $f \sim f_2 + g$ where f_2 is hyperbolic. Moreover, f represents all elements of k.*

This is a translation of prop. 3 and its corollary.

Corollary 1.—*Let $g = g(X_1, \ldots, X_{n-1})$ be a nondegenerate quadratic form and let $a \in k^*$. The following properties are equivalent:*

(i) *g represents a.*
(ii) *One has $g \sim h + aZ^2$ where h is a form in $n-2$ variables.*
(iii) *The form $f = g \dot{-} aZ^2$ represents 0.*

It is clear that (ii) $\Rightarrow$ (i). Conversely, if g represents a, the quadratic module V corresponding to g contains an element x such that $x.x = a$; if H denotes the orthogonal complement to x, we have $V = H \hat{\oplus} kx$, hence $g \sim h + aZ^2$ where h denotes the quadratic form attached to a basis of H.

The implication (ii) $\Rightarrow$ (iii) is immediate. Finally, if the form $f = g \dot{-} aZ^2$ has a nontrivial zero $(x_1, \ldots, x_{n-1}, z)$ we have either $z = 0$ in which case g represents 0 thus also a, or $z \neq 0$ in which case $g(x_1/z, \ldots, x_{n-1}/z) = a$. Hence (iii) $\Rightarrow$ (i).

Corollary 2.—*Let g and h be two nondegenerate forms of rank $\geqq 1$, and let $f = g \dot{-} h$. The following properties are equivalent:*

(a) *f represents 0.*
(b) *There exists $a \in k^*$ which is represented by g and by h.*
(c) *There exists $a \in k^*$ such that $g \dot{-} aZ^2$ and $h \dot{-} aZ^2$ represent 0.*

The equivalence (b) $\Leftrightarrow$ (c) follows from corollary 1. The implication (b) $\Rightarrow$ (a) is trivial. Let us show (a) $\Rightarrow$ (b). A nontrivial zero of f can be written in the form (x, y) with $g(x) = h(y)$. If the element $a = g(x) = h(y)$ is $\neq 0$, it is clear that (b) is verified. If $a = 0$, one of the forms g for example, represents 0, thus all elements of k, and in particular all non-zero values taken by h.

Theorem 1 translates into the classical decomposition of quadratic forms into "sums of squares":

Theorem 1'.—*Let f be a quadratic form in n variables. There exists $a_1, \ldots, a_n \in k$ such that $f \sim a_1 X_1^2 + \ldots + a_n X_n^2$.*

The *rank* of f is the number of indices i such that $a_i \neq 0$. It is equal to n if and only if the *discriminant* $a_1 \ldots a_n$ of f is $\neq 0$ (in other words, if f is nondegenerate).

Finally the corollary to Witt's theorem gives the following "cancellation theorem":

Theorem 4.—*Let $f = g + h$ and $f' = g' + h'$ be two nondegenerate quadratic forms. If $f \sim f'$ and $g \sim g'$, one has $h \sim h'$.*

Corollary.—*If f is nondegenerate, then*

$$f \sim g_1 + \ldots + g_m + h$$

where $g_1, \ldots, g_m$ are hyperbolic and h does not represent 0. This decomposition is unique up to equivalence.

The existence follows from prop. 3', and the uniqueness from theorem 4.

[The number m of hyperbolic factors can be characterized as the dimension of the *maximal isotropic subspaces* of the quadratic module defined by f.]

1.7. *Quadratic forms over* $\mathbf{F}_q$

Let p be a prime number $\neq 2$ and let $q = p^f$ a power of p; let $\mathbf{F}_q$ be a field with q elements (cf. chap. I, §1).

Proposition 4.—*A quadratic form over $\mathbf{F}_q$ of rank $\geqq 2$ (resp. of rank $\geqq 3$) represents all elements of $\mathbf{F}_q^*$ (resp. of $\mathbf{F}_q$).*

In view of cor. 1 of prop. 3', it suffices to prove that all quadratic forms in 3 variables represent 0 and this has been proved in chap. I, §2, as a consequence of Chevalley theorem.

[Let us indicate how one can prove this proposition without using Chevalley theorem. One has to show that, if $a, b, c \in \mathbf{F}_q$ are not zero, the equation

$$(*) \quad ax^2 + by^2 = c$$

has a solution. Let A (resp. B) be the set of elements of $\mathbf{F}_q$ of the form ax^2 (resp. of the form $c - by^2$) with $x \in \mathbf{F}_q$ (resp. with $y \in \mathbf{F}_q$). One sees easily that A and B have each $(q+1)/2$ elements; thus $A \cap B \neq \varnothing$ from which one gets a solution of $(*)$.]

Recall now that the group $\mathbf{F}_q^*/\mathbf{F}_q^{*2}$ has two elements (chap. I, n° 3.1). Let a denote an element of $\mathbf{F}_q^*$ which is not a square.

Proposition 5.—*Every nondegenerate quadratic form of rank n over $\mathbf{F}_q$ is equivalent to*

$$X_1^2 + \ldots + X_{n-1}^2 + X_n^2$$

or

$$X_1^2 + \ldots + X_{n-1}^2 + aX_n^2,$$

depending on whether its discriminant is a square or not.

This is clear if $n = 1$. If $n \geqq 2$, prop. 4 shows that the form f represents 1. It is thus equivalent to X_1^2+g where g is a form in $n-1$ variables and one applies the inductive hypothesis to g.

Corollary.—*For two nondegenerate quadratic forms over* $\mathbf{F}_q$ *to be equivalent it is necessary and sufficient that they have same rank and same discriminant.*

(Of course the discriminant is viewed as an element of the quotient group $\mathbf{F}_q^*/\mathbf{F}_q^{*2}$.)

§2. *Quadratic forms over* $\mathbf{Q}_p$

In this paragraph (n° 2.4 excepted) p is a prime number and k is the p-adic field $\mathbf{Q}_p$.

All quadratic modules are over k and nondegenerate; we make the same conventions for the quadratic forms.

2.1. *The two invariants*

Let (V, Q) be a quadratic module of rank n and $d(Q)$ its discriminant; it is an element of k^*/k^{*2}, cf. n° 1.1. If $\boldsymbol{e} = (e_1, \ldots e_n)$ is an orthogonal basis of V, and if we put $a_i = e_i.e_i$, we have

$$d(Q) = a_1 \ldots a_n \quad (\text{in } k^*/k^{*2})$$

(In what follows we will often denote by the same letter an element of k^* and its class modulo k^{*2}.)

Recall now that, if a and b are elements of k^*, we have defined in chap. III, n° 1.1, the Hilbert symbol (a, b), equal to ± 1. We put

$$\varepsilon(\boldsymbol{e}) = \prod_{i<j} (a_i, a_j).$$

One has $\varepsilon(\boldsymbol{e}) = \pm 1$. Moreover $\varepsilon(\boldsymbol{e})$ is an *invariant* of (V, Q):

Theorem 5.—*The number* $\varepsilon(\boldsymbol{e})$ *does not depend on the choice of the orthogonal basis* $\boldsymbol{e}$.

If $n = 1$, one has $\varepsilon(\boldsymbol{e}) = 1$. If $n = 2$, one has $\varepsilon(\boldsymbol{e}) = 1$ if and only if the form $Z^2-a_1X^2-a_2Y^2$ represents 0, that is to say (cf. cor. 1 to prop. 3') if and only if $a_1X^2+a_2Y^2$ represents 1; but this last condition signifies that there exists $v \in V$ such that $Q(v) = 1$ and this does not depend on $\boldsymbol{e}$. For $n \geqq 3$ we use induction on n. By th. 2 it suffices to prove that $\varepsilon(\boldsymbol{e}) = \varepsilon(\boldsymbol{e}')$ when $\boldsymbol{e}$ and $\boldsymbol{e}'$ are *contiguous*. In view of the symmetry of the Hilbert symbol, $\varepsilon(\boldsymbol{e})$ does not change when we permute the e_1; we can thus suppose that $\boldsymbol{e}' = (e_1' \ldots, e_n')$ is such that $e_1' = e_1$. If we put $a_i' = e_i'.e_i'$, then $a_1' = a_1$. One can write $\varepsilon(\boldsymbol{e})$ in the form

$$\begin{aligned}\varepsilon(\boldsymbol{e}) &= (a_1, a_2 \ldots a_n) \prod_{2 \leqq i<j} (a_i, a_j) \\ &= (a_1, d(Q)a_1) \prod_{2 \leqq i<j} (a_i, a_j)\end{aligned}$$

since $d(Q) = a_1 \ldots a_n$.

Similarly

$$\varepsilon(e') = (a_1, d(Q)a_1) \prod_{2 \leq i < j} (a'_i, a'_j).$$

But the inductive hypothesis applied to the orthogonal complement of e_1 shows that

$$\prod_{2 \leq i < j} (a_i, a_j) = \prod_{2 \leq i < j} (a'_i, a'_j),$$

from which the desired result follows.

We write from now on $\varepsilon(Q)$ instead of $\varepsilon(e)$.

Translation.—If f is a quadratic form in n variables and if

$$f \sim a_1 X_1^2 + \ldots + a_n X_n^2,$$

the two elements

$$d(f) = a_1 \ldots a_n \quad (\text{in } k^*/k^{*2})$$

$$\varepsilon(f) = \prod_{i<j} (a_i, a_j) \quad (\text{in } \{\pm 1\})$$

are *invariants* of the equivalence class of f.

2.2. *Representation of an element of k by a quadratic form*

Lemma.—a) *The number of elements in the $\mathbf{F}_2$-vector space k^*/k^{*2} is 2^r with $r = 2$ if $p \neq 2$ and $r = 3$ if $p = 2$.*

b) *If $a \in k^*/k^{*2}$ and $\varepsilon = \pm 1$, let H_a^ε be the set of $x \in k^*/k^{*2}$ such that $(x, a) = \varepsilon$. If $a = 1$, H_a^1 has 2^r elements and $H_a^{-1} = \varnothing$. If $a \neq 1$, H_a^ε has 2^{r-1} elements.*

c) *Let $a, a' \in k^*/k^{*2}$ and $\varepsilon, \varepsilon' = \pm 1$; assume that H_a^ε and $H_{a'}^{\varepsilon'}$ are nonempty. For $H_a^\varepsilon \cap H_{a'}^{\varepsilon'} = \varnothing$, it is necessary and sufficient that $a = a'$ and $\varepsilon = -\varepsilon'$.*

Assertion a) has been proved in chap. II, n° 3.3. In b) the case $a = 1$ is trivial; of $a \neq 1$, the homomorphism $b \mapsto (a, b)$ carries k^*/k^{*2} *onto* $\{\pm 1\}$ (chap. III, n° 1.2, th. 2); its kernel H_a^1 is thus a hyperplane of k^*/k^{*2} and has 2^{r-1} elements; its complement H_a^{-1} has 2^{r-1} elements (it is an "affine" hyperplane parallel to H_a^1). Finally, if H_a^ε and $H_{a'}^{\varepsilon'}$ are nonempty and disjoint, they have necessarily 2^{r-1} elements each and are complementary to one another; this implies $H_a^1 = H_{a'}^1$ hence

$$(x, a) = (x, a') \quad \text{for all } x \in k^*/k^{*2};$$

since the Hilbert symbol is nondegenerate, this implies $a = a'$ and $\varepsilon = -\varepsilon'$. The converse is trivial.

Let now f be a quadratic form of rank n; let $d = d(f)$ and $\varepsilon = \varepsilon(f)$ be its two invariants.

Theorem 6.—*For f to represent 0 it is necessary and sufficient that:*

i) $n = 2$ *and* $d = -1$ (in k^*/k^{*2}),

ii) $n = 3$ *and* $(-1, -d) = \varepsilon$,

iii) $n = 4$ *and either* $d \neq 1$ *or* $d = 1$ *and* $\varepsilon = (-1, -1)$.
iv) $n \geqq 5$
(In particular, *all forms in at least 5 variables represent* 0.)

Before proving the theorem, let us indicate a consequence of it: let $a \in k^*/k^{*2}$ and $f_a = f \dot{-} aZ^2$. We know (cf. n° 1.6) that f_a represents 0 if and only if f represents a. On the other hand,

$$d(f_a) = -ad, \quad \varepsilon(f_a) = (-a, d)\varepsilon,$$

as one checks right away. By applying theorem 6 to f_a, and taking into account the above formulas, we obtain:

Corollary. *Let $a \in k^*/k^{*2}$. In order that f represent a it is necessary and sufficient that:*

i) $n = 1$ *and* $a = d$,
ii) $n = 2$ *and* $(a, -d) = \varepsilon$,
iii) $n = 3$ *and either* $a \neq -d$ *or* $a = -d$ *and* $(-1, -d) = \varepsilon$,
iv) $n \geqq 4$.

(Note that, in this statement as in theorem 6, a and d are viewed as elements of k^*/k^{*2}; the inequality $a \neq -d$ means that a is not equal to the product of $-d$ by a square.)

Proof of theorem 6.—We write f in the form $f \sim a_1X_1^2 + \ldots + a_nX_n^2$ and consider separately the cases $n = 2, 3, 4$ and $\geqq 5$.

i) *The case* $n = 2$.
The form f represents 0 if and only if $-a_1/a_2$ is a square; but $-a_1/a_2 = -a_1a_2 = -d$ in k^*/k^{*2}; hence this means that $d = -1$.

ii) *The case* $n = 3$.
The form f represents 0 if and only if the form

$$-a_3f \sim -a_3a_1X_1^2 - a_3a_2X_2^2 - X_3^2$$

represents 0. Now by the very definition of the Hilbert symbol, this last form represents 0 if and only if we have

$$(-a_3a_1, -a_3a_2) = 1.$$

Expanding this, we find:

$$(-1, -1)(-1, a_1)(-1, a_2)(a_3, a_3)(a_1, a_2)(a_1, a_3)(a_2, a_3) = 1$$

But one has $(a_3, a_3) = (-1, a_3)$, cf. chap. III, n° 1.1, prop. 2, formula iv). One can thus rewrite the above condition in the form

$$(-1, -1)(-1, a_1a_2a_3)(a_1, a_2)(a_1, a_3)(a_2, a_3) = 1,$$

or $(-1, -d)\varepsilon = 1$, i.e. $(-1, -d) = \varepsilon$.

iii) *The case* $n = 4$.
By cor. 2 to prop. 3′, f represents 0 if and only if there exists an element $x \in k^*/k^{*2}$ which is represented by the two forms

$$a_1X_1^2+a_2X_2^2 \qquad \text{and } -a_3X_3^2-a_4X_4^2.$$

By case ii) of the corollary above, such an x is characterized by the conditions

$$(x, -a_1a_2) = (a_1, a_2) \qquad \text{and} \qquad (x, -a_3a_4) = (-a_3, -a_4).$$

Let A be the subset of k^*/k^{*2} defined by the first condition, and let B be the subset defined by the second. In order that f does not represent 0, it is necessary and sufficient that $A \cap B = \varnothing$. Now A and B are clearly nonempty (one has $a_1 \in A$ and $-a_3 \in B$ for example). By part c) of the lemma given at the beginning of this n°, the relation $A \cap B = \varnothing$ is thus equivalent to

$$a_1a_2 = a_3a_4 \qquad \text{and} \qquad (a_1, a_2) = -(-a_3 - a_4).$$

The first condition means that $d = 1$. If it is fulfilled one has

$$\varepsilon = (a_1, a_2)(a_3, a_4)(a_3a_4, a_3a_4);$$

by using the relation $(x, x) = (-1, x)$ (cf. chap. III, n° 1.1, formula iv) of prop. 2) we get from this:

$$\begin{aligned}\varepsilon &= (a_1, a_2)(a_3, a_4)(-1, a_3a_4)\\ &= (a_1, a_2)(-a_3, a_4)(-1, -1).\end{aligned}$$

Hence the second condition can be written $\varepsilon = -(-1, -1)$, from which the result follows.

iv) *The case* $n \geqq 5$.

It is sufficient to treat the case $n = 5$. By using the lemma and part ii) of the above corollary, we see that a form of rank 2 represents at least 2^{r-1} elements of k^*/k^{*2}, and, the same is *a fortiori* true for the form of rank $\geqq 2$. Since $2^{r-1} \geqq 2$, f represents at least one element $a \in k^*/k^{*2}$ *distinct* from d. One has

$$f \sim aX^2 + g,$$

where g is a form of rank 4. The discriminant of g is equal to d/a; it is thus different from 1, and, by iii), the form g represents 0. The same is then true for f, and the proof of theorem 6 is complete.

Remarks.—1) Let f be a quadratic form not representing 0. The above results show that the number of elements of k^*/k^{*2} which are represented by f is equal to 1 if $n = 1$, to 2^{r-1} if $n = 2$, to $2^r - 1$ if $n = 3$, and to 2^r if $n = 4$.

2) We have seen that all quadratic forms in 5 variables over $\mathbf{Q}_p$ represent 0. In this connection, let us mention a conjecture made by E. Artin: *all homogeneous polynomials of degree d over* $\mathbf{Q}_p$ *in at least* d^2+1 *variables have a nontrivial zero.* The case $d = 3$ has been solved affirmatively (see, for example, T. Springer, *Koninkl. Nederl. Akad. van Wetenss.*, 1955, pp. 512–516). The general case remained open for about thirty years. It was only in 1966 that G. Terjanian showed that *Artin's conjecture is false*: there exists a homogeneous polynomial of degree 4 over $\mathbf{Q}_2$ in 18 variables which has no nontrivial zero. Terjanian starts from the polynomial

$$n(X, Y, Z) = X^2YZ+Y^2ZX+Z^2XY+X^2Y^2+Y^2Z^2+Z^2X^2-X^4-Y^4-Z^4,$$

which has the property that $n(x, y, z) \equiv -1 \pmod 4$ if (x, y, z) is primitive in $(\mathbf{Z}_2)^3$. Let

$$f(X_1, \ldots, X_9) = n(X_1, X_2, X_3) + n(X_4, X_5, X_6) + n(X_7, X_8, X_9);$$

one has $f(x_1, \ldots, x_9) \not\equiv 0 \pmod 4$ if $(x_1, \ldots, x_9)$ is primitive. From this one deduces easily that the polynomial

$$F(X_1, \ldots, X_{18}) = f(X_1, \ldots, X_9)+4f(X_{10}, \ldots, X_{18})$$

does not have a non-trivial zero. (There exist analogous examples—but of higher degrees—for all the $\mathbf{Q}_p$.)

One knows nevertheless that Artin's conjecture is "almost" true: for a fixed degree d, it holds for all prime numbers p except a finite number (AX-KOCHEN, *Amer. J. of Math.*, 1965); however, even for $d = 4$, one does not know how to determine the set of exceptional prime numbers.

2.3. *Classification*

Theorem 7.—*Two quadratic forms over k are equivalent if and only if they have the same rank, same discriminant, and same invariant* ε.

That two equivalent forms have the same invariants follows from the definitions. The converse is proved by induction on the rank n of the two forms f and g considered (the case $n = 0$ being trivial). Corollary to theorem 6 shows that f and g represent the same elements of k^*/k^{*2}; one can thus find $a \in k^*$ which is represented at the same time by f and by g; this allows one to write:

$$f \sim aZ^2+f' \text{ and } g \sim aZ^2+g',$$

where f', g' are forms of rank $n-1$. One has

$$d(f') = ad(f) = ad(g) = d(g')$$

$$\varepsilon(f') = \varepsilon(f)\,(a, d(f')) = \varepsilon(g)\,(a, d(g')) = \varepsilon(g'),$$

which shows that f' and g' have the same invariants. In view of the inductive hypothesis, we have $f' \sim g'$, hence $f \sim g$.

Corollary.—*Up to equivalence, there exists a unique form of rank* 4 *which does not represent* 0; *if* $(a, b) = -1$, *it is the form* $z^2-ax^2-by^2+abt^2$.

Indeed, by th. 6, such a form is characterized by

$$d(f) = 1, \varepsilon(f) = -(-1,-1)$$

and a simple computation shows that $z^2-ax^2-by^2+abt^2$ has these properties.

Remark.—This form is the *reduced norm* of the unique non-commutative field of degree 4 over $\mathbf{Q}_p$; this field can be defined as a field "of quaternions" with basis $\{1, i, j, k\}$ where $i^2 = a, j^2 = b, ij = k = -ji$, and $(a, b) = -1$.

Proposition 6.—*Let $n \geqq 1$, $d \in k^*/k^{*2}$ and $\varepsilon = \pm 1$. In order that there exists a quadratic form f of rank n such that $d(f) = d$ and $\varepsilon(f) = \varepsilon$ it is necessary and sufficient that $n = 1, \varepsilon = 1$; or $n = 2, d \neq -1$; or $n = 2, \varepsilon = 1$; or $n \geqq 3$.*

The case $n = 1$ is trivial. If $n = 2$, one has $f \sim aX^2 + bY^2$ and,

$$\text{if } d(f) = -1, \text{ then } \varepsilon(f) = (a, b) = (a, -ab) = 1;$$

thus we cannot have simultaneously $d(f) = -1$ and $\varepsilon(f) = -1$. Conversely, if $d = -1, \varepsilon = 1$, we take $f = X^2 - Y^2$; if $d \neq -1$, there exists $a \in k^*$ such that $(a, -d) = \varepsilon$ and we take $f = aX^2 + adY^2$.

If $n = 3$, we choose $a \in k^*/k^{*2}$ distinct from $-d$; by what we have just seen, there exists a form g of rank 2 such that $d(g) = ad$, $\varepsilon(g) = \varepsilon(a, -d)$; the form $aZ^2 + g$ works. The case $n \geqq 4$ is reduced to the case $n = 3$ by taking $f = g(X_1, X_2, X_3) + X_4^2 + \ldots + X_n^2$ where g has the required invariants.

Corollary.—*The number of classes of quadratic forms of rank n over $\mathbf{Q}_p$ for $p \neq 2$ (resp. $p = 2$) is equal to 4 (resp. 8) if $n = 1$, to 7 (resp. 15) if $n = 2$, and to 8 (resp. 16) if $n \geqq 3$.*

Indeed, $d(f)$ can take 4 (resp. 8) values and $\varepsilon(f)$ can take 2 values.

2.4. *The real case*

Let f be a quadratic form of rank n over the field $\mathbf{R}$ of real numbers. We know that f is equivalent to

$$X_1^2 + \ldots + X_r^2 - Y_1^2 - \ldots - Y_s^2,$$

where r and s are two integers $\geqq 0$ such that $r + s = n$; the pair (r, s) depends only on f; it is called the *signature* of f. We say that f is *definite* if r or $s = 0$, i.e. if f does not change sign; otherwise, we say that f is *indefinite*. (This is the case where f represents 0.)

The *invariant* $\varepsilon(f)$ is defined as in the case of $\mathbf{Q}_p$; due to the fact that $(-1, -1) = -1$, we have

$$\varepsilon(f) = (-1)^{s(s-1)/2} = \begin{cases} 1 \text{ if } s \equiv 0, 1 \pmod 4 \\ -1 \text{ if } s \equiv 2, 3 \pmod 4. \end{cases}$$

Moreover:

$$d(f) = (-1)^s = \begin{cases} 1 \text{ if } s \equiv 0 \pmod 2 \\ -1 \text{ if } s \equiv 1 \pmod 2. \end{cases}$$

We see that the knowledge of $d(f)$ and $\varepsilon(f)$ is equivalent to that of the class of s modulo 4; in particular, *$d(f)$ and $\varepsilon(f)$ determine f up to equivalence if $n \leqq 3$.*

One also checks that *parts* i), ii), iii) *of th.* 6 *and its corollary are valid for* $\mathbf{R}$ (indeed their proofs use only the nondegeneracy of the Hilbert symbol, and this applies to $\mathbf{R}$); it is clear that part iv) does not extend.

§3. *Quadratic forms over* Q

All quadratic forms considered below have coefficients in $\mathbf{Q}$ and are nondegenerate.

3.1. *Invariants of a form*

As in chap. III, §2, we denote by V the union of the set of prime numbers and the symbol ∞, and we put $\mathbf{Q}_\infty = \mathbf{R}$.

Let $f \sim a_1X_1^2 + \ldots + a_nX_n^2$ be a quadratic form of rank n. We associate to it the following invariants:

a) The discriminant $d(f) \in \mathbf{Q}^*/\mathbf{Q}^{*2}$ equal to $a_1 \ldots a_n$.
b) Let $v \in V$. The injection $\mathbf{Q} \to \mathbf{Q}_v$ allows one to view f as a quadratic form (which we will denote f_v) over $\mathbf{Q}_v$. The invariants of f_v will be denoted by $d_v(f)$ and $\varepsilon_v(f)$; it is clear that $d_v(f)$ is the image of $d(f)$ by $\mathbf{Q}^*/\mathbf{Q}^{*2} \to \mathbf{Q}_v^*/\mathbf{Q}_v^{*2}$; we have

$$\varepsilon_v(f) = \prod_{i<j} (a_i, a_j)_v.$$

The product formula (chap. III, n° 2.1, th. 3) gives the relation

$$\prod_{v \in V} \varepsilon_v(f) = 1.$$

c) The *signature* (r, s) of the real quadratic form f is another invariant of f.

The invariants $d_v(f)$, $\varepsilon_v(f)$, and (r, s) are sometimes called the *local* invariants of f.

3.2. *Representation of a number by a form*

Theorem 8 (Hasse-Minkowski).—*In order that f represent* 0, *it is necessary and sufficient that, for all $v \in V$, the form f_v represent* 0.

(In other words: f has a "global" zero if and only if f has everywhere a "local" zero.)

The necessity is trivial. In order to see the sufficiency, we write f in the form

$$f = a_1X_1^2 + \ldots + a_nX_n^2, \quad a_i \in \mathbf{Q}^*.$$

Replacing f by $a_1 f$, one can moreover suppose that $a_1 = 1$. We consider separately the cases $n = 2$, 3, 4 and $\geqq 5$.

i) *The case $n = 2$.*

We have $f = X_1^2 - aX_2^2$; since f_∞ represents 0, a is > 0. If we write a in the form

$$a = \prod_p p^{v_p(a)},$$

the fact that f_p represents 0 shows that a is a square in $\mathbf{Q}_p$, hence that $v_p(a)$ is *even*. From this follows that a is a square in $\mathbf{Q}$ and f represents 0.

ii) *The case* $n = 3$ (Legendre).

We have $f = X_1^2 - aX_2^2 - bX_3^2$; being free to multiply a, b, by squares, we can assume that a and b are *square free integers* (i.e. $v_p(a)$, $v_p(b)$ are equal to 0 or 1 for all prime numbers p). Also we can assume that $|a| \leqq |b|$. We use induction on the integer $m = |a|+|b|$. If $m = 2$, we have

$$f = X_1^2 \pm X_2^2 \pm X_3^2;$$

the case of $X_1^2+X_2^2+X_3^2$ is excluded because f_∞ represents 0; in the other cases, f represents zero.

Suppose now that $m > 2$, i.e. $|b| \geqq 2$ and write b in the form

$$b = \pm p_1 \ldots p_k,$$

where the p_i are distinct primes. Let p be one of the p_i; we are going to prove that *a is a square modulo p*. This is obvious if $a \equiv 0 \pmod p$. Otherwise a is a p-adic unit; by hypothesis, there exists $(x, y, z) \in (\mathbf{Q}_p)^3$ such that $z^2 - ax^2 - by^2 = 0$ and we can suppose that (x, y, z) is primitive (cf. chap. II, n° 2.1, prop. 6). We have $z^2 - ax^2 \equiv 0 \pmod p$. From this follows that, if $x \equiv 0 \pmod p$, the same is true also for z, and by^2 is divisible by p^2; since $v_p(b) = 1$ this implies $y \equiv 0 \pmod p$ contrary to the fact that (x, y, z) is primitive. Thus we have $x \not\equiv 0 \pmod p$, which shows that a is a square (mod p). Now, since $\mathbf{Z}/b\mathbf{Z} = \Pi\mathbf{Z}/p_i\mathbf{Z}$, we see that *$a$ is a square modulo b*. There exist thus integers t, b' such that

$$t^2 = a + bb'$$

and we can choose t in such a way that $|t| \leqq |b|/2$. The formula $bb' = t^2 - a$ shows that bb' is a *norm* of the extension $k(\sqrt{a})/k$ where $k = \mathbf{Q}$ or $\mathbf{Q}_v$; from this we conclude (the argument is the same as that for prop. 1 of chap. III), that f represents 0 in k if and only if the same is true for

$$f' = X_1^2 - aX_2^2 - b'X_3^2.$$

In particular, f' represents 0 in each of the $\mathbf{Q}_v$. But we have:

$$|b'| = \left|\frac{t^2-a}{b}\right| \leqq \frac{|b|}{4} + 1 < |b| \qquad \text{because } |b| \geqq 2.$$

Write b' in the form $b''u^2$ with b'', u integers and b'' square free; we have *a fortiori* $|b''| < |b|$. The induction hypothesis applies thus to the form $f'' = X_1^2 - aX_2^2 - b''X_3^2$ which is equivalent to f'; hence this form represents 0 in $\mathbf{Q}$ and the same is true for f.

iii) *The case* $n = 4$.

Write $f = aX_1^2 + bX_2^2 - (cX_3^2 + dX_4^2)$. Let $v \in V$. Since f_v represents 0, cor. 2 of prop. 3′ of n° 1.6 shows that there exists $x_v \in \mathbf{Q}_v^*$ which is represented both by $aX_1^2 + bX_2^2$ and by $cX_3^2 + dX_4^2$; by part ii) to cor. of th. 6 (which applies equally well to $\mathbf{Q}_\infty = \mathbf{R}$), this is equivalent to saying that

$$(x_v, -ab)_v = (a, b)_v \quad \text{and} \quad (x_v, -cd)_v = (c, d)_v \text{ for all } v \in V.$$

Since $\prod_{v \in V} (a, b)_v = \prod_{v \in V} (c, d)_v = 1$, we can apply th. 4 of chap. III, n° 2.2 and obtain from it the existence of $x \in \mathbf{Q}^*$ such that

$$(x, -ab)_v = (a, b)_v \quad \text{and} \quad (x, -cd)_v = (c, d)_v \text{ for all } v \in V.$$

The form $aX_1^2+bX_2^2-xZ^2$ represents zero in each of the $\mathbf{Q}_v$ hence in $\mathbf{Q}$ by what we have just seen. Hence x is represented in $\mathbf{Q}$ by $aX_1^2+bX_2^2$, and the same argument applies to $cX_3^2+dX_4^2$; the fact that f represents 0 follows from this.

iv) *The case* $n \geqq 5$.

We use induction on n. We write f in the form

$$f = h \dot{-} g$$

with $h = a_1X_1^2+a_2X_2^2$, $g = -(a_3X_3^2+\ldots+a_nX_n^2)$.

Let S be the subset of V consisting of ∞, 2, and the numbers p such that $v_p(a_i) \neq 0$ for one $i \geqq 3$; it is a finite set. Let $v \in S$. Since f_v represents 0, there exists $a_v \in \mathbf{Q}_v^*$ which is represented in $\mathbf{Q}_v$ by h and by g; there exist $x_i^v \in \mathbf{Q}_v$, $i = 1, \ldots, n$, such that

$$h(x_1^v, x_2^v) = a_v = g(x_3^v, \ldots, x_n^v).$$

But the squares of $\mathbf{Q}_v^*$ form an *open* set (cf. chap. II, n° 3.3). Using the approximation theorem (chap. III, n° 2.2, lemma 2), this implies the existence of $x_1, x_2 \in \mathbf{Q}$ such that, if $a = h(x_1, x_2)$, one has $a/a_v \in \mathbf{Q}_v^{*2}$ for all $v \in S$. Consider now the form $f_1 = aZ^2 \dot{-} g$. If $v \in S$, g represents a_v in $\mathbf{Q}_v$, thus also a because $a/a_v \in \mathbf{Q}_v^{*2}$; hence f_1 represents 0 in $\mathbf{Q}_v$. If $v \notin S$, the coefficients $-a_3, \ldots, -a_n$ of g are v-adic units; the same is true of $d_v(g)$, and because $v \neq 2$ we have $\varepsilon_v(g) = 1$. [This could also be deduced from cor. 2 to th. 1 of chap. II, n° 2.2. combined with Chevalley's theorem.] In all cases, we see that f_1 represents 0 in $\mathbf{Q}_v$; since the rank of f_1 is $n-1$, the inductive hypothesis shows that f_1 represents 0 in $\mathbf{Q}$, i.e. g represents a in $\mathbf{Q}$; since h represents a, f represents 0, and the proof is complete.

Corollary 1.—*Let* $a \in \mathbf{Q}^*$. *In order that* f *represents* a *in* $\mathbf{Q}$ *it is necessary and sufficient that it does in each of the* $\mathbf{Q}_v$.

This follows from the theorem applied to the form $aZ^2 \dot{-} f$.

Corollary 2 (Meyer).—*A quadratic form of rank* $\geqq 5$ *represents* 0 *if and only if it is indefinite* (i.e. if it represents 0 in $\mathbf{R}$).

Indeed, by th. 6, such a form represents 0 in each of the $\mathbf{Q}_p$.

Corollary 3.—*Let* n *be the rank of* f. *Suppose that* $n = 3$ (*resp.* $n = 4$ *and* $d(f) = 1$.) *If* f *represents* 0 *in all the* $\mathbf{Q}_v$ *except at most one, then* f *represents* 0.

Suppose that $n = 3$. By th. 6, f represents 0 in $\mathbf{Q}_v$ if and only if we have:

$$(*)_v \quad (-1, -d(f))_v = \varepsilon_v(f).$$

But the two families $\varepsilon_v(f)$, $(-1, -d(f))_v$ satisfy *the product formula* of chap.

III, n° 2.1. From this follows that, if $(*)_v$ is true for all v except at most one $(*)_v$ is true for all v; by th. 8, f represents 0.

When $n = 4$ and $d(f) = 1$ we argue in the same way, the equality $(*)_v$ being replaced by $(-1, -1)_v = \varepsilon_v(f)$.

Remarks.—1) Suppose that $n = 2$ and that f represents 0 in all the $\mathbf{Q}_v$ *except a finite number*. One can then show, by means of the theorem on arithmetic progressions (cf. chap. VI, n° 4.3) that f represents 0.

2) Th. 8 does not extend to homogeneous polynomials of degree ≥ 3; for example, Selmer has proved that the equation

$$3X^3+4Y^3+5Z^3 = 0$$

has a nontrivial solution in each of the $\mathbf{Q}_v$ but none in $\mathbf{Q}$.

3.3. *Classification*

Theorem 9.—*Let f and f' be two quadratic forms over $\mathbf{Q}$. For f and f' to be equivalent over $\mathbf{Q}$ it is necessary and sufficient that they are equivalent over each $\mathbf{Q}_v$.*

The necessity is trivial. To prove the sufficiency, we use induction on the rank n of f and f'. If $n = 0$, there is nothing to prove. Otherwise, there exists $a \in \mathbf{Q}^*$ represented by f, thus also by f' (cf. cor. 1 to th. 8). Thus we have $f \sim aZ^2+g$, $f' \sim aZ^2+g'$. By th. 4 of n° 1.6, we have $g \sim g'$ over $\mathbf{Q}_v$ for all $v \in V$. The induction hypothesis then shows that $g \sim g'$ over $\mathbf{Q}$, hence $f \sim f'$.

Corollary.—*Let (r, s) and (r', s') be the signatures of f and of f'. For f and f' to be equivalent it is necessary and sufficient that one has*

$$d(f) = d(f'), \quad (r, s) = (r', s'), \quad \text{and } \varepsilon_v(f) = \varepsilon_v(f') \text{ for all } v \in V.$$

Indeed these conditions just mean that f and f' are equivalent over each of the $\mathbf{Q}_v$.

Remark.—The invariants $d = d(f)$, $\varepsilon_v = \varepsilon_v(f)$ and (r, s) are not arbitrary. They verify the following relations:

(1) $\varepsilon_v = 1$ for almost all $v \in V$ and $\prod_{v \in V} \varepsilon_v = 1$,

(2) $\varepsilon_v = 1$ if $n = 1$ or if $n = 2$ and if the image d_v of d in $\mathbf{Q}_v^*/\mathbf{Q}_v^{*2}$ is equal to -1,

(3) $r, s \geqq 0$ and $r+s = n$,

(4) $d_\infty = (-1)^s$,

(5) $\varepsilon_\infty = (-1)^{s(s-1)/2}$

Conversely:

Proposition 7.—*Let d, $(\varepsilon_v)_{v \in V}$, and (r, s) verify the relations* (1) *to* (5) *above. Then there exists a quadratic form of rank n over $\mathbf{Q}$ having for invariants d, $(\varepsilon_v)_{v \in V}$ and (r, s).*

The case $n = 1$ is trivial.

Suppose that $n = 2$. Let $v \in V$. The nondegeneracy of the Hilbert symbol, together with condition (2), shows that there exists $x_v \in \mathbf{Q}_v^*$ such that $(x_v, -d)_n = \varepsilon_v$. From this and condition (1), follows the existence of $x \in \mathbf{Q}^*$ such that $(x, -d)_v = \varepsilon_v$ for all $v \in V$ (chap. III, n° 2.2, th. 4). The form xX^2+xdY^2 works.

Suppose that $n = 3$. Let S be the set of $v \in V$ such that $(-d, -1)_v = -\varepsilon_v$; it is a finite set. If $v \in S$, choose in $\mathbf{Q}_v^*/\mathbf{Q}_v^{*2}$ an element c_v distinct from the image $-d_v$ of $-d$ in this group. Using the approximation theorem (chap. III, n° 2.2, lemma 2), we see that there exists $c \in \mathbf{Q}^*$ whose image in each of the $\mathbf{Q}_v^*/\mathbf{Q}_v^{*2}$, $v \in S$, is c_v. From what we have just proved follows the existence of a form g of rank 2 such that

$$d(g) = cd, \quad \varepsilon_v(g) = (c, -d)_v\varepsilon_v \quad \text{for all } v \in V.$$

The form $f = cZ^2+g$ then works. [Note that for $n \leqq 3$ we do not need to consider the signature of the form, since conditions (3), (4), (5) determine it as a function of d_∞ and ε_∞].

When $n \geqq 4$ we use induction on n. Suppose first that $r \geqq 1$. Using the inductive hypothesis, we obtain a form g of rank $n-1$ which has for invariants d, $(\varepsilon_v)_{v \in V}$ and $(r-1, s)$; the form X^2+g works. When $r = 0$, we use a form h of rank $n-1$ having for invariants $-d$, $\varepsilon_v(-1, -d)_v$, and $(0, n-1)$; the form $-X^2+h$ works.

Appendix

Sums of three squares

Let n and p be positive integers. We say that *n is the sum of p squares* if n is representable *over the ring* $\mathbf{Z}$ by the quadratic form $X_1^2+\ldots+X_p^2$, i.e. if there exist integers $n_1, \ldots, n_p$ such that

$$n = n_1^2+\ldots+n_p^2$$

Theorem (Gauss).—*In order that a positive integer be a sum of three squares it is necessary and sufficient that it is not of the form* $4^a(8b-1)$ *with* $a, b \in \mathbf{Z}$.

(Example: if n is not divisible by 4, it is a sum of three squares if and only if $n \equiv 1, 2, 3, 5, 6$ (mod. 8).)

Proof.—We can suppose n is nonzero. The condition "n is of the form $4^a(8b-1)$" is then equivalent to say that *$-n$ is a square in* $\mathbf{Q}_2^*$ (chap. II, n° 3.3, th. 4). But we have:

Lemma A.—*Let $a \in \mathbf{Q}^*$. In order that a be represented in $\mathbf{Q}$ by the form $f = X_1^2+X_2^2+X_3^2$ it is necessary and sufficient that a is >0 and that $-a$ is not a square in* $\mathbf{Q}_2$.

By cor. 1 of th. 8 we have to express that a is represented by f in $\mathbf{R}$ and in all $\mathbf{Q}_p$. The case of $\mathbf{R}$ gives the positivity condition. On the other hand the local invariants $d_p(f)$ and $\varepsilon_p(f)$ are equal to 1. If $p \neq 2$, one has

$$(-1, -d_p(f))_p = (-1, -1)_p = 1 = \varepsilon_p(f);$$

the corollary of theorem 6 thus shows that a is represented by f in $\mathbf{Q}_p$. If $p = 2$, we have $(-1, -d_2(f))_2 = -1 \neq \varepsilon_2(f)$; the same corollary shows that a is represented by f in $\mathbf{Q}_2$ if and only if a is different from -1 in $\mathbf{Q}_2^*/\mathbf{Q}_2^{*2}$, i.e. if $-a$ is not a square in $\mathbf{Q}_2$.

Now we must pass from representations *in* $\mathbf{Q}$ to representations *in* $\mathbf{Z}$. This is done by means of the following lemma:

Lemma B (Davenport-Cassels).—*Let* $f(X) = \sum_{i,j=1}^{p} a_{ij}X_iX_j$ *be a positive definite quadratic form, the matrix* (a_{ij}) *being symmetric and with integer coefficients. We make the following hypothesis:*

(H) For every $x = (x_1, \ldots, x_p) \in \mathbf{Q}^p$ *there exists* $y \in \mathbf{Z}^p$ *such that* $f(x-y) < 1$.

If $n \in \mathbf{Z}$ *is represented by* f *in* $\mathbf{Q}$, *then* n *is also represented by* f *in* $\mathbf{Z}$.

If $x = (x_1, \ldots, x_p)$ and $y = (y_1, \ldots, y_p)$ are two elements of $\mathbf{Q}^p$, we denote by $x.y$ their scalar product $\Sigma a_{ij}x_iy_j$. One has $x.x = f(x)$.

Let n be an integer represented by f in $\mathbf{Q}$. There exists an integer $t > 0$ such that $t^2n = x.x$ with $x \in \mathbf{Z}^p$. Choose t and x in such a way that t is minimum; we must prove that $t = 1$.

By hypothesis (H), there exists $y \in \mathbf{Z}^p$ such that

$$\frac{x}{t} = y+z \qquad \text{with } z.z < 1.$$

If $z.z = 0$ we have $z = 0$ and $\frac{x}{t}$ has integer coefficients. Because of the minimality of t, this implies $t = 1$.

Assume now that $z.z \neq 0$ and put

$$\begin{aligned} a &= y.y-n \\ b &= 2(nt-x.y) \\ t' &= at+b \\ x' &= ax+by. \end{aligned}$$

One has $a, b, t' \in \mathbf{Z}$, and:

$$\begin{aligned} x'.x' &= a^2x.x+2abx.y+b^2y.y \\ &= a^2t^2n+ab(2nt-b)+b^2(n+a) \\ &= n(a^2t^2+2abt+b^2) \\ &= t'^2n. \end{aligned}$$

Moreover:

$$\begin{aligned} tt' &= at^2+bt = t^2y.y-nt^2+2nt^2-2t\,x.y \\ &= t^2y.y-2t\,x.y+x.x = (ty-x).(ty-x) \\ &= t^2z.z, \end{aligned}$$

hence $t' = t\, z.z$; since $0 < z.z < 1$, we have $0 < t' < t$. This contradicts the minimality of t and concludes the proof of the lemma.

To prove the theorem, it suffices now to check that the form $f = X_1^2 + X_2^2 + X_3^2$ satisfies condition (H) of lemma B. But this is clear: if $(x_1, x_2, x_3) \in \mathbf{Q}^3$ we choose $(y_1, y_2, y_3) \in \mathbf{Z}^3$ such that $|x_i - y_i| \leqq \frac{1}{2}$ for all i; we have $\Sigma(x_i - y_i)^2 \leqq 3/4 < 1$.

Corollary 1 (Lagrange).—*Every positive integer is a sum of four squares.*

Let n be an integer >0. We write n in the form $4^a m$, where m is not divisible by 4. If $m \equiv 1, 2, 3, 5, 6 \pmod 8$, m is a sum of three squares, and the same holds for n. Otherwise $m \equiv -1 \pmod 8$ and $m-1$ is sum of three squares; in this case m is a sum of four squares, and the same holds for n.

Corollary 2 (Gauss).—*Every positive integer is a sum of three triangular numbers.*

(A number is called "triangular" if it is of the form $m(m+1)/2$ with $m \in \mathbf{Z}$.)

Let n be a positive integer. By applying the theorem to $8n+3$, we see that there exist integers x_1, x_2, x_3 such that

$$x_1^2 + x_2^2 + x_3^2 = 8n+3.$$

One has

$$x_1^2 + x_2^2 + x_3^2 \equiv 3 \pmod 8.$$

But the only squares in $\mathbf{Z}/8\mathbf{Z}$ are 0, 1 and 4; a sum of three squares in $\mathbf{Z}/8\mathbf{Z}$ can be equal to 3 only if each of its terms equal 1. This shows that the x_i are odd, and one can write them in the form $2m_i + 1$ with m_i an integer. We have

$$\sum_{i=1}^{i=3} \frac{m_i(m_i+1)}{2} = \frac{1}{8}\left(\sum_{i=1}^{i=3} (2m_i+1)^2 - 3\right) = \frac{1}{8}(8n+3-3) = n.$$

Chapter V

Integral Quadratic Forms with Discriminant ± 1

§1. *Preliminaries*

1.1. *Definitions*

Let n be an integer $\geqq 0$. We are interested in the following category S_n:

An object E of S_n is a *free abelian group of rank* n (i.e. isomorphic to $\mathbf{Z}^n$) together with a symmetric bilinear form $E \times E \to \mathbf{Z}$, denoted by $(x, y) \mapsto x.y$, such that:

(i) *The homomorphism of* E *into* $\operatorname{Hom}(E, \mathbf{Z})$ *defined by the form* $x.y$ *is an isomorphism.*

One sees easily that this condition is equivalent to the following (cf. Bourbaki, *Alg.*, chap. IX, §2, prop. 3):

(ii) *If* (e_i) *is a base of* E, *and if* $a_{ij} = e_i.e_j$, *the determinant of the matrix* $A = (a_{ij})$ *is equal to* ± 1.

The notion of *isomorphism* of two objects $E, E' \in S_n$ is defined in an obvious way. One then writes $E \simeq E'$. It is also convenient to introduce $S = \cup S_n$, $n = 0, 1, \ldots$

If $E \in S_n$, the function $x \mapsto x.x$ makes E a *quadratic module* over $\mathbf{Z}$ (cf. chap. IV, def. 1, n° 1.1). If (e_i) is a basis of E and if $x = \Sigma x_i e_i$, the quadratic form $f(x) = x.x$ is given by the formula

$$f(x) = \sum_{i,j} a_{ij} x_i x_j \qquad \text{with } a_{ij} = e_i.e_j$$

$$= \sum_i a_{ii} x_i^2 + 2\sum_{i<j} a_{ij} x_i x_j.$$

The coefficients of the non-diagonal terms are thus even. The discriminant of f (i.e. $\det(a_{ij})$) is equal to ± 1. Changing the basis (e_i) means replacing the matrix $A = (a_{ij})$ by ${}^tBAB$ with $B \in \mathbf{GL}(n, \mathbf{Z})$. From the point of view of the form f, this means changing the variables (x_i) by the linear substitution with matrix B; the form so obtained is called *equivalent* to the form f. (Observe that this is an equivalence *over the ring* $\mathbf{Z}$ *of integers*; it is *finer* than the equivalence over $\mathbf{Q}$ studied in the preceding chapter.)

1.2. *Operations on* S

Let $E, E' \in S$. We denote by $E \oplus E'$ the *direct sum* of E and of E' together with the bilinear form which is the direct sum of those on E and E'; by definition (cf. Bourbaki, *Alg.*, chap. IX, § 1, n° 3):

$$(x+x').(y+y') = x.y+x'.y' \qquad \text{if } x, y \in E \text{ and } x', y' \in E'.$$

From the point of view of "quadratic forms" this operation corresponds to that of *orthogonal direct sum* denoted by $\hat{\oplus}$ in chap. IV.

One can also define *tensor products* $E \otimes E'$ and *exterior powers* $\wedge^m E$ (BOURBAKI, *loc cit.*, n° 9); we will not need them.

1.3. *Invariants*

1.3.1. If $E \in S_n$, the integer n is called the *rank* of E and is denoted by $r(E)$.

1.3.2. Let $E \in S$ and let $V = E \otimes \mathbf{R}$ the $\mathbf{R}$-vector space obtained by extending the scalars from $\mathbf{Z}$ to $\mathbf{R}$. The quadratic form of V has a well defined signature (r, s) (chap. IV, n° 2.4). The integer

$$\tau(E) = r - s$$

is called the *index* of E. One has

$$-r(E) \leqq \tau(E) \leqq r(E) \quad \text{and} \quad r(E) \equiv \tau(E) \pmod 2.$$

Recall that E is called *definite* if $\tau(E) = \pm r(E)$, i.e. if $x.x$ has constant sign; otherwise E is called *indefinite*.

1.3.3. The *discriminant* of E with respect to a basis (e_i) does not depend on the choice of this basis. Indeed, changing the basis (e_i) multiplies the discriminant by $\det(X^t X) = \det(X)^2$ where X is an invertible matrix over $\mathbf{Z}$; the determinant of X is equal to ± 1, and its square is equal to 1.

The discriminant of E is denoted by $d(E)$; one has $d(E) = \pm 1$. If $V = E \otimes \mathbf{R}$ is of signature (r, s), the sign of $d(E)$ is $(-1)^s$; since $d(E) = \pm 1$, we get the formula:

$$d(E) = (-1)^{(r(E) - \tau(E))/2}.$$

1.3.4. Let $E \in S$. We say that E is *even* (or of *type* II) if the quadratic form associated with E takes only even values; if A is the matrix defined by a basis of E, this amounts to saying that *all the diagonal terms of A are even.*

If E is not even, we say that E is *odd* (or of *type* I).

1.3.5. Let $E \in S$ and let $\bar{E} = E/2E$ be the reduction of E modulo 2. It is a vector space of dimension $r(E)$ over the field $\mathbf{F}_2 = \mathbf{Z}/2\mathbf{Z}$. By passage to quotient, the form $x.y$ defines on $\bar{E}$ a form $\bar{x}.\bar{y}$ which is symmetric and of discriminant $\pm 1 = 1$. The associated quadratic form $\bar{x}.\bar{x}$ is additive:

$$(\bar{x}+\bar{y}).(\bar{x}+\bar{y}) = \bar{x}.\bar{x}+\bar{y}.\bar{y}+2\bar{x}.\bar{y} = \bar{x}.\bar{x}+\bar{y}.\bar{y}$$

Thus it is an element of the *dual* of $\bar{E}$. But the bilinear form $\bar{x}.\bar{y}$ is non-degenerate; it defines an isomorphism of $\bar{E}$ on its dual. From this we see that there exists a *canonical element* $\bar{u} \in \bar{E}$ such that

$$\bar{u}.\bar{x} = \bar{x}.\bar{x} \quad \text{for all } \bar{x} \in \bar{E}.$$

Lifting $\bar{u}$ to E, we obtain an element $u \in E$, unique modulo $2E$, such that

$$u.x \equiv x.x \pmod 2 \qquad \text{for all } x \in E.$$

Consider the integer $u.u$. If we replace u by $u+2x$, $u.u$ is replaced by

$$(u+2x).(u+2x) = u.u+4(u.x+x.x) \equiv u.u \pmod 8.$$

The image of $u.u$ in $\mathbf{Z}/8\mathbf{Z}$ is thus an *invariant* of E; we denote it by $\sigma(E)$. If E is of type II, the form $\bar{x}.\bar{x}$ is zero (in other words, $x.y$ is *alternating*) and we can take $u = 0$, hence $\sigma(E) = 0$.

1.3.6. Let p be a prime number, and let $V_p = E \otimes \mathbf{Q}_p$ be the $\mathbf{Q}_p$-vector space obtained from E by extending the scalars from $\mathbf{Z}$ to $\mathbf{Q}_p$. The invariant $\varepsilon(V_p) = \pm 1$ of V_p defined in chap. IV, n° 2.1 is *a fortiori* an invariant of E; we denote it by $\varepsilon_p(E)$. One can prove that:

$$\varepsilon_p(E) = 1 \text{ if } p \neq 2$$

$$\varepsilon_2(E) = (-1)^j, \text{ where } j = \tfrac{1}{8}(d(E)+r(E)-\sigma(E)-1).$$

This is seen by splitting $E \otimes \mathbf{Z}_p$ in an orthogonal direct sum of $\mathbf{Z}_p$-modules of rank 1 (resp. rank 1 or 2) if $p \neq 2$ (resp. $p = 2$). Since we do not use these formulas, we leave the details of the verification to the reader (see also J. Cassels, *Comm. Math. Helv.*, 37, 1962, pp. 61–64).

1.3.7. Let E_1, $E_2 \in S$ and $E = E_1 \oplus E_2$. In order that E is of type II it is necessary and sufficient that E_1 and E_2 are of type II. One has:

$$r(E) = r(E_1)+r(E_2), \quad \tau(E) = \tau(E_1)+\tau(E_2)$$

$$\sigma(E) = \sigma(E_1)+\sigma(E_2), \quad d(E) = d(E_1).d(E_2).$$

1.4. *Examples*

1.4.1. We denote by I_+ (resp. I_-) the $\mathbf{Z}$-module $\mathbf{Z}$ with the bilinear form xy (resp. $-xy$); the corresponding quadratic form is $+x^2$ (resp. $-x^2$).

If s and t are two positive integers, we denote by $sI_+ \oplus tI_-$ the direct sum of s copies of I_+ and t copies of I_-; the corresponding quadratic form is $\sum_{i=1}^{s} x_i^2 - \sum_{j=1}^{t} y_j^2$. The invariants of this module are:

$$r = s+t, \ \tau = s-t, \ d = (-1)^t, \ \sigma \equiv s-t \pmod 8.$$

Aside the trivial case $(s, t) = (0, 0)$, the module $sI_+ \oplus tI_-$ is of type I.

1.4.2. We denote by U the element of S_2 defined by the matrix $\begin{pmatrix} 0 & 1 \\ 1 & 0 \end{pmatrix}$. The associated quadratic form is $2x_1x_2$; U is of type II. One has:

$$r(U) = 2, \ \tau(U) = 0, \ d(U) = -1, \ \sigma(U) = 0$$

1.4.3. Let k be a positive integer, let $n = 4k$, and let V be the vector space $\mathbf{Q}^n$ with the standard bilinear form $\Sigma x_i y_i$ corresponding to the unit matrix. Let $E_0 = \mathbf{Z}^n$ be the subgroup of V formed from the points with integer coordinates, endowed with the bilinear form induced from that of V; E_0 is an element of S_n isomorphic to nI_+. Let E_1 be the submodule of E_0

formed of elements x such that $x.x \equiv 0 \pmod 2$, i.e. $\Sigma x_i \equiv 0 \pmod 2$. One has $(E_0:E_1) = 2$. Let E be the submodule of V generated by E_1 and by $e = (\frac{1}{2}, \ldots, \frac{1}{2})$. One has $2e \in E_1$ (since $n \equiv 0 \pmod 4$) and $e \notin E_1$, hence $(E:E_1) = 2$. For an element $x = (x_i)$ of V to belong to E, it is necessary and sufficient that one has

$$2x_i \in \mathbf{Z}, \quad x_i - x_j \in \mathbf{Z}, \quad \sum_{i=1}^{n} x_i \in 2\mathbf{Z}.$$

Then we have $x.e = \frac{1}{2}\Sigma x_i \in \mathbf{Z}$; since $e.e = k$, this shows that the form $x.y$ *takes integral values on E*. Moreover, the fact that E_1 has the same index in E_0 and in E shows that the discriminant of E is equal to that of E_0, that is to say $+1$. The quadratic module E is thus an element of $S_n = S_{4k}$; we denote it by Γ_n. When k is *even* (i.e. when $n \equiv 0 \pmod 8$) $e.e = k$ is even and this implies that $x.x$ is even for all $x \in E$; Γ_n *is thus of type* II *when* $n \equiv 0 \pmod 8$. One has

$$r(\Gamma_{8m}) = 8m, \quad \tau(\Gamma_{8m}) = 8m, \quad \sigma(\Gamma_{8m}) = 0, \quad d(\Gamma_{8m}) = 1.$$

The case of Γ_8 is particularly interesting. There are 240 elements[1] $x \in \Gamma_8$ such that $x.x = 2$; if (e_i) denotes the canonical base of $\mathbf{Q}^8$, these are the vectors

$$\pm e_i \pm e_k \, (i \neq k) \quad \text{and} \quad \tfrac{1}{2}\sum_{i=1}^{8} \varepsilon_i e_i, \; \varepsilon_i = \pm 1, \prod_{i=1}^{8} \varepsilon_i = 1.$$

[The mutual scalar products of these vectors are integers; they form what is called in Lie group theory a "*root system of type* E_8" see, BOURBAKI, *Gr. et Alg. de Lie*, chap. VI, §4, n° 10.]

One can take as a basis of Γ_8 the elements

$$\tfrac{1}{2}(e_1+e_8) - \tfrac{1}{2}(e_2+\ldots+e_7), \; e_1+e_2, \quad \text{and} \quad e_i - e_{i-1} \, (2 \leqq i \leqq 7).$$

The corresponding matrix is

$$\begin{pmatrix} 2 & 0 & -1 & 0 & 0 & 0 & 0 & 0 \\ 0 & 2 & 0 & -1 & 0 & 0 & 0 & 0 \\ -1 & 0 & 2 & -1 & 0 & 0 & 0 & 0 \\ 0 & -1 & -1 & 2 & -1 & 0 & 0 & 0 \\ 0 & 0 & 0 & -1 & 2 & -1 & 0 & 0 \\ 0 & 0 & 0 & 0 & -1 & 2 & -1 & 0 \\ 0 & 0 & 0 & 0 & 0 & -1 & 2 & -1 \\ 0 & 0 & 0 & 0 & 0 & 0 & -1 & 2 \end{pmatrix}$$

For $m \geqq 2$ the vectors $x \in \Gamma_{8m}$ such that $x.x = 2$ are simply the vectors $\pm e_i \pm e_k (i \neq k)$; note that they do not generate Γ_{8m}, contrary to what happens for $m = 1$. In particular, $\Gamma_8 \oplus \Gamma_8$ is not isomorphic to Γ_{16}.

---

[1] More generally we will show in chap. VII, n° 6.5, that, if N is an integer $\geqq 1$, the number of $x \in \Gamma_8$ such that $x.x = 2N$ is equal to 240 times the sum of the cubes of divisors of N.

1.5. *The group $K(S)$*

Let E, $E' \in S$. We say that E and E' are *stably isomorphic* if there exists $F \in S$ such that $E \oplus F \simeq E' \oplus F$; this is an equivalence relation. We denote by $K_+(S)$ the quotient of S by this relation and if $E \in S$, we denote by (E) the class of E in $K_+(S)$. The operation $\oplus$ defines by passage to quotient a composition law $+$ on $K_+(S)$; this law is commutative, associative, and has for neutral element the class 0 of the module $0 \in S$. One has

$$(E \oplus E') = (E)+(E').$$

Moreover, if x, y, $z \in K_+(S)$ are such that $x+z = y+z$, one has $x = y$; the proof is immediate. This allows us to define the *group* $K(S)$ from the semi-group $K_+(S)$ (exactly as one defines $\mathbf{Z}$ from the set $\mathbf{Z}_+$ of positive integers): By definition, an element of $K(S)$ is a pair (x, y) with $x, y \in K_+(S)$, two pairs (x, y), (x', y') being identified if and only if $x+y' = y+x'$. The composition law of $K(S)$ is defined by

$$(x, y)+(x', y') = (x+x', y+y').$$

It makes $K(S)$ into a commutative group with neutral element (0, 0). We identify $K_+(S)$ with a subset of $K(S)$ by the map $x \mapsto (x, 0)$. Every element of $K(S)$ is a difference of two elements of $K_+(S)$, thus can be written in the form $(E)-(F)$ with E, $F \in S$. One has

$$(E)-(F) = (E')-(F') \quad \text{in } K(S)$$

if and only if there exists $G \in S$ such that $E \oplus F' \oplus G \simeq E' \oplus F \oplus G$, i.e. if and only if $E \oplus F'$ and $E' \oplus F$ are stably isomorphic.

Universal property of $K(S)$.—Let A be a commutative group and let $f\colon S \to A$ be a function such that $f(E) = f(E_1)+f(E_2)$ if $E \simeq E_1 \oplus E_2$; we then say that f is *additive*. If $x = (E)-(F)$ is an element of $K(S)$ we put

$$f(X) = f(E)-f(F);$$

this does not depend on the chosen decomposition of x. It is obvious that the function $f\colon K(S) \to A$ thus defined is a *homomorphism*. Conversely, every homomorphism $f\colon K(S) \to A$ gives, by composition with $S \to K(S)$, an additive function on S. We express this "universal" property of $K(S)$ by saying that $K(S)$ is the *Grothendieck group* of S relative to the operation $\oplus$.

In particular, the invariants r, τ, d, σ of n° 1.3 define homomorphisms

$$r\colon K(S) \to \mathbf{Z}, \quad \tau\colon K(S) \to \mathbf{Z}, \quad d\colon K(S) \to \{\pm 1\}, \quad \sigma\colon K(S) \to \mathbf{Z}/8\mathbf{Z}.$$

We have again $\tau \equiv r \bmod 2$ and $d = (-1)^{(r-\tau)/2}$.

§2. *Statement of results*

2.1. *Determination of the group $K(S)$*

Theorem 1.—*The group $K(S)$ is a free abelian group with basis (I_+) and (I_-).* (The proof will be given in n° 3.4.)

In other terms, all $f \in K(S)$ can be written uniquely as $f = s.(I_+)+t.(I_-)$ with $s, t \in \mathbf{Z}$. One has $r(f) = s+t$, $\tau(f) = s-t$, which shows that s and t are determined by r and τ. From this follows:

Corollary 1.—*The pair* (r, τ) *defines an isomorphism of* $K(S)$ *onto the subgroup of* $\mathbf{Z}\times\mathbf{Z}$ *formed of elements* (a, b) *such that* $a \equiv b \pmod 2$.
Hence:

Corollary 2.—*For two elements E and E' of S to be stably isomorphic it is necessary and sufficient that they have same rank and same index.*

[Note that this does not imply $E \simeq E'$ For example $U = \begin{pmatrix} 0 & 1 \\ 1 & 0 \end{pmatrix}$ defines in $K(S)$ the same element as $\begin{pmatrix} 1 & 0 \\ 0 & -1 \end{pmatrix} = I_+ \oplus I_-$, but U and $I_+ \oplus I_-$ are of different types.]

Theorem 2.—*One has* $\sigma(E) \equiv \tau(E) \pmod 8$ *for every* $E \in S$.
Indeed τ reduced mod 8, and σ, are homomorphisms of $K(S)$ in $\mathbf{Z}/8\mathbf{Z}$ which are equal on the generators I_+ and I_- of $K(S)$; hence they coincide.

Corollary 1.—*If E is of type* II, *one has* $\tau(E) \equiv 0 \pmod 8$.
Indeed $\sigma(E) = 0$.
(Note that this implies that $r(E) \equiv 0 \pmod 2$ and $d(E) = (-1)^{r(E)/2}$.)

Corollary 2.—*If E is definite and of type* II, *one has* $r(E) \equiv 0 \pmod 8$.
Indeed we have $\tau(E) = \pm r(E)$ if E is definite.

Remarks—1) Conversely, we saw in n° 1.4 that for all n divisible by 8, there exists $E \in S_n$ which is positive definite and of type II.
2) The congruence $\sigma(E) \equiv \tau(E) \pmod 8$ can also be deduced from the product formula $\Pi\varepsilon_v(E) = 1$ (see chap. IV, n° 3.1) combined with the values of $\varepsilon_p(E)$ given (without proof) in n° 1.3.6.

2.2. *Structure theorems (indefinite case)*

Let $E \in S$. One says that E *represents zero* if there exists $x \in E$, $x \neq 0$, such that $x.x = 0$. This is equivalent to saying that the corresponding quadratic form $Q(x)$ *represents* 0 *over* $\mathbf{Q}$ in the sense of chap. IV, n° 1.6; indeed, one passes from a rational zero to an integral zero by a homothety.

Theorem 3.—*If $E \in S$ is indefinite, E represents zero.*
(The proof will be given in n° 3.1.)

Theorem 4.—*If $E \in S$ is indefinite and of type* I, *E is isomorphic to $sI_+ \oplus tI_-$ where s and t are integers* $\geqq 1$.
[The corresponding quadratic form is thus equivalent over $\mathbf{Z}$ to the form $\sum_{i=1}^{s} x_i^2 - \sum_{j=1}^{t} x_j^2$.]
(The proof will be given in n° 3.3.)

Corollary.—*Let E and E' be two elements of S with the same rank and index. Then either $E \oplus I_+ \simeq E' \oplus I_+$ or $E \oplus I_- \simeq E' \oplus I_-$.*

This is clear if $E = 0$. Otherwise, one of the two modules $E \oplus I_+$ or $E \oplus I_-$ is indefinite. Suppose that the first is. Since E and E' have the same signature, $E' \oplus I_+$ is also indefinite. By applying theorem 4, we see that $E \oplus I_+$ and $E' \oplus I_+$ are isomorphic to $sI_+ \oplus tI_-$ and $s'I_+ \oplus t'I_-$ respectively. Since E and E' have the same signature, we have $s = s'$ and $t = t'$, hence the result.

Theorem 5.—*If $E \in S$ is indefinite of type* II, *and if $\tau(E) \geqq 0$, then E is isomorphic to $pU \oplus q\Gamma_8$ where p and q are positive integers.*

[When $\tau(E) \leqq 0$, one has a corresponding result obtained by applying the theorem to the module-E deduced from E by changing the sign of the quadratic form.]

(The proof will be given in n° 3.5.)

Note that $q = \frac{1}{8}\tau(E)$ and $p = \frac{1}{2}(r(E) - \tau(E))$. This shows that E is determined up to isomorphism by its rank and its index. Since the same is true for type I (cf. theorem 4), we get:

Theorem 6.—*If E, $E' \in S$ are indefinite, and have same rank, index, and type, they are isomorphic.*

2.3. *The definite case*

One does not have a structure theorem, but only a *finiteness theorem*: for all integers n, S_n contains but a finite number of positive definite classes. This follows, for instance, from the "reduction theory" of quadratic forms. The explicit determination of these classes has been made only for small values of n (for $n \leqq 16$, see M. KNESER, *Archiv der Math.*, 8, 1957, pp. 241–250). One can get this from the *Minkowski-Siegel formula* (Kneser uses a different method). I will just state this formula (I restrict myself, for the sake of simplicity, to type II—there are analogous results for type I):

Let $n = 8k$ be an integer divisible by 8. Let C_n denote the set of isomorphism classes of elements $E \in S_n$ which are positive definite of type II. If $E \in C_n$, let G_E be the *group of automorphisms* of E; it is a finite group since it is a discrete subgroup of the orthogonal group, which is compact; let g_E be the order of G_E. Put:

$$M_n = \sum_{E \in C_n} 1/g_E.$$

This is the "mass" of C_n, in Eisenstein's sense, i.e. the number of elements E of C_n, counted each with the multiplicity $1/g_E$. The Minkowski-Siegel formula[1] gives the value of M_n:

$$(*)\ M_n = \frac{B_{2k}}{8k} \prod_{j=1}^{j=4k-1} \frac{B_j}{4j},$$

[1] For a proof of this formula, cf. C. L. SIEGEL, *Gesamm. Abh.*, I, n° 20 and III, n° 79.

where $n = 8k$, and the B_j are the Bernoulli numbers $\left(B_1 = \frac{1}{6}, B_2 = \frac{1}{30}, \ldots,\right.$ cf. chap. VII, n° 4.1).

(Here are some approximate values of the M_n:
$M_8 = 10^{-9} \times 1.4352\ldots$; $M_{16} = 10^{-18} \times 2.4885\ldots$; $M_{24} = 10^{-15} \times 7.9367\ldots$, $M_{32} = 10^{7} \times 4.0309\ldots$; $M_{40} = 10^{51} \times 4.3930\ldots$)

This formula gives a method to prove that a subset C' of C_n is *equal* to C_n; it suffices to check that the sum of the $1/g_E$, for $E \in C'$, is equal to M_n (if $C' \neq C_n$, this sum is $< M_n$).

Examples

i) $n = 8$, i.e. $k = 1$. One can show (see for instance BOURBAKI, *Gr. et Alg. de Lie*, chap. VI, §4, n° 10) that the order of the group of automorphisms of Γ_8 is $2^{14}3^5 5^2 7$. Moreover, formula (*) gives $M_8 = 2^{-14}\,3^{-5}\,5^{-2}\,7^{-1}$. By comparing, we see that C_8 is *reduced to the single element* Γ_8, a result due to Mordell.

ii) $n = 16$. We know two elements of C_{16}: Γ_{16} and $\Gamma_8 \oplus \Gamma_8$. One can prove that the corresponding orders g_E are respectively $2^{15}(16!)$ and $2^{29}3^{10}5^4 7^2$. Moreover $M_{16} = 691.2^{-30}3^{-10}5^{-4}7^{-2}11^{-1}13^{-1}$ and it is easy to check that

$$691/2^{30}3^{10}5^4 7^2 11.13 = 1/2^{15}(16!) + 1/2^{29}3^{10}5^4 7^2.$$

We thus have $C_{16} = \{\Gamma_{16}, \Gamma_8 \oplus \Gamma_8\}$, a result due to Witt.

iii) $n = 24$. The determination of C_{24} has been made in 1968 by H. Niemeier[1]; this set has 24 elements. One of them (discovered by Leech à propos of the sphere-packing problem in $\mathbf{R}^{24}$) is particularly remarkable; it is the only element of C_{24} which contains no vector x with $x.x = 2$. Its group of automorphisms G has order:

$$2^{22}3^9 5^4 7^2 11.13.23 = 8\,315\,553\,613\,086\,720\,000.$$

The quotient $G/\{\pm 1\}$ is the new simple group discovered by Conway[2].

iv) $n = 32$. Since $M_{32} > 4.10^7$ and $g_E \geqq 2$ for all E, we see that C_{32} has more than 80 million elements; they have not been classified yet.

§3. *Proofs*

3.1. *Proof of theorem* 3

Let $E \in S_n$ and let $V = E \otimes \mathbf{Q}$ the corresponding $\mathbf{Q}$-vector space. Suppose E is indefinite. We must show that E (or V) represents zero. We consider several cases:

[1] See H. NIEMEIER, *J. Number Theory*, 5, 1973, pp. 142–178.

[2] See J. H. CONWAY, *Proc. Nat. Acad. Sci. USA*, 61, 1968, pp. 398–400, and *Invent. Math.*, 7, 1969, pp. 137–142.

i) $n = 2$. The signature of V is then $(1, 1)$, hence $d(E) = -1$. Since $-d(E)$ is a square in $\mathbf{Q}$, it is clear that V represents 0.

ii) $n = 3$. Let $f(X_1, X_2, X_3) = \Sigma a_{ij}X_iX_j$ be the corresponding quadratic form with respect to a basis of E; one has $a_{ij} \in \mathbf{Z}$ and $\det(a_{ij}) = \pm 1$. If p is a prime number $\neq 2$, the form deduced from f by reduction modulo p has a nontrivial zero (chap. I, n° 2.2), and this zero can be lifted to a p-adic zero (chap. II, n° 2.2), and cor. 2 to th. 1). Hence f represents 0 in all the $\mathbf{Q}_p(p \neq 2)$, and also in $\mathbf{R}$; by cor. 3 of th. 8 of chap IV, n° 3.2, this implies that f represents 0 in $\mathbf{Q}$.

iii) $n = 4$. The same argument as above shows that the quadratic form f represents 0 in all the $\mathbf{Q}_p$, $p \neq 2$, and in $\mathbf{R}$. If the discriminant $d(E)$ of f is equal to 1, this suffices to imply that f represents 0 in $\mathbf{Q}$ (cor. 3 to th. 8 of chap. IV, n° 3.2). Otherwise one has $d(E) = -1$ and $d(E)$ is not a square in $\mathbf{Q}_2$; by th. 6 of n° 2.2. of chap. IV, this implies that f represents 0 in $\mathbf{Q}_2$; the Hasse-Minkowski's theorem (chap. IV, n° 3.2, th. 8) then shows that f represents 0 in $\mathbf{Q}$.

iv) $n \geqq 5$. One applies Meyer's theorem (chap. IV, n° 3.2, cor. 2 to th. 8).

3.2. *Lemmas*

Let $E \in S$ and let F be a submodule of E; let F' be the set of elements of E orthogonal to all elements of F.

Lemma 1.—*In order that F, with the form $x.y$ induced from that of E, be in S it is necessary and sufficient that E be the direct sum of F and F'.*

If $E = F \oplus F'$, then one has $d(E) = d(F).d(F')$ from which $d(F') = \pm 1$. Conversely if $d(F) = \pm 1$, one has clearly $F \cap F' = 0$; moreover, if $x \in E$, the linear form $y \mapsto x.y (y \in F)$ is defined by an element $x_0 \in F$. We then have $x = x_0 + x_1$, with $x_0 \in F$ and $x_1 \in F'$, hence $E = F \oplus F'$.

Lemma 2.—*Let $x \in E$ be such that $x.x = \pm 1$ and let X be the orthogonal complement of x in E. If $D = \mathbf{Z}x$, one has $E = D \oplus X$.*

One applies lemma 1 to $F = D$. (If, for instance, $x.x = 1$, one has $D \simeq I_+$ hence $E \simeq I_+ \oplus X$.)

An element $x \in E$ is called *indivisible* if it is not contained in any subgroup nE ($n \geqq 2$), i.e., if one cannot divide it by any integer $\geqq 2$. Every non-zero element of E can be written in a unique way in the form mx with $m \geqq 1$ and x indivisible.

Lemma 3.—*If x is an indivisible element of E there exists $y \in E$ such that $x.y = 1$.*

Let f_x be the linear form $y \mapsto x.y$ defined by x. It is a homomorphism $E \to \mathbf{Z}$. Moreover, f_x is *indivisible* since x is and since $x.y$ defines an isomorphism of E onto its dual $\mathrm{Hom}(E, \mathbf{Z})$. From this follows that f_x is surjective (otherwise, one could divide it by an integer $\geqq 2$) and there exists thus $y \in E$ such that $x.y = 1$.

3.3. *Structure theorem (odd indefinite case*[1]*)*

Lemma 4.—*Let $E \in S_n$. Suppose E is indefinite and of type* I. *There exists $F \in S_{n-2}$ such that $E \sim I_+ \oplus I_- \oplus F$.*

By theorem 3, there exists $x \in E$, $x \neq 0$ such that $x.x = 0$. Being free to divide x by an integer, we can suppose x is indivisible; by lemma 3 above, there exists thus $y \in E$ such that $x.y = 1$. One can choose y of such that $y.y$ is *odd.* Indeed, suppose that $y.y$ is even; since E is of type I, there exists $t \in E$ such that $t.t$ is odd. Put $y' = t+ky$ and choose k such that $x.y' = 1$, i.e. $k = 1-x.t$; one has $y'.y' \equiv t.t \pmod 2$ and $y'.y'$ is odd. We can thus suppose that $y.y = 2m+1$. Put then $e_1 = y-mx$, $e_2 = y-(m+1)x$. One checks immediately that $e_1.e_1 = 1$, $e_1.e_2 = 0$, $e_2.e_2 = -1$. The submodule G of E generated by (e_1, e_2) is isomorphic to $I_+ \oplus I_-$; by lemma 1, we have thus $E \simeq I_+ \oplus I_- \oplus F$, with $F \in S_{n-2}$.

Proof of theorem 4.—We use induction on n. Let $E \in S_n$ with E indefinite and of type I. By lemma 4, $E \simeq I_+ \oplus I_- \oplus F$. If $n = 2$, we have $F = 0$ and the theorem is proved. If $n > 2$, we have $F \neq 0$ and one of the modules $I_+ \oplus F$, $I_- \oplus F$, is indefinite, for instance the first one. Since I_+ is of type I, the same is true for $I_+ \oplus F$ and the inductive hypothesis shows that $I_+ \oplus F$ is of the form $aI_+ \oplus bI_-$; this shows that $E \simeq aI_+ \oplus (b+1)I_-$.

3.4. *Determination of the group $K(S)$*

Let $E \in S$, $E \neq 0$. Then $E \oplus I_+$ or $E \oplus I_-$ is indefinite and of type I. Applying theorem 4, we see that the image of E in $K(S)$ is a linear combination of (I_+) and of (I_-). This implies that (I_+) and (I_-) generate $K(S)$. Since their images by the homomorphism

$$(r, \tau): K(S) \to \mathbf{Z} \times \mathbf{Z}$$

are linearly independent, (I_+) and (I_-) form a basis of $K(S)$.

3.5. *Structure theorem (even indefinite case)*

Lemma 5.—*Let $E \in S$. Suppose E is indefinite and of type* II. *There exists $F \in S$ such that $E \simeq U \oplus F$.*

We proceed as in the proof of lemma 4. Choose first $x \in E$, $x \neq 0$, x indivisible such that $x.x = 0$; choose next $y \in E$ such that $x.y = 1$. If $y.y = 2m$, we replace y by $y-mx$ and obtain a new y such that $y.y = 0$. The submodule G of E generated by (x, y) is then isomorphic to U; by lemma 1 one has $E \simeq U \oplus F$ with $F \in S$.

Lemma 6.—*Let $F_1, F_2 \in S$. Suppose that F_1 and F_2 are of type* II *and that $I_+ \oplus I_- \oplus F_1 \simeq I_+ \oplus I_- \oplus F_2$. Then $U \oplus F_1 \simeq U \oplus F_2$.*

---

[1] The method followed in this section has been shown to me by Milnor, together with the idea of introducing the group $K(S)$.

To simplify the notations, we put $W = I_+ \oplus I_-$, $E_i = W \oplus F_i$, and $V_i = E_i \otimes \mathbf{Q}$. In E_i, let E_i^0 be the subgroup of elements x such that $x.x \equiv 0 \pmod 2$; it is of index 2 in E_i. One sees immediately that $E_i^0 = W^0 \oplus F_i$ where W^0 is the set of elements $x = (x_1, x_2)$ of W such that $x_1 \equiv x_2 \pmod 2$. Let E_i^+ be the "dual" of E_i^0 in V_i, i.e. the set of $y \in V_i$ such that $x.y \in \mathbf{Z}$ for all $x \in E_i^0$. It is clear that $E_i^+ = W^+ \oplus F_i$ where W^+ is the set of (x_1, x_2) such that $2x_1 \in \mathbf{Z}$, $2x_2 \in \mathbf{Z}$, $x_1 - x_2 \in \mathbf{Z}$. One has $E_i^0 \subset E_i \subset E_i^+$ and the quotient E_i^+/E_i^0 is isomorphic to W^+/W_0; it is a group of type (2, 2). There exist thus three subgroups strictly between E_1^0 and E_i^+; they correspond to the three subgroups of order 2 in a group of type (2, 2). One of them is E_i itself; the two others will be denoted by E_i' and E_i''. Here again we have:

$$E_i' = W' \oplus F_i \quad \text{and} \quad E_i'' = W'' \oplus F_i$$

where W' and W'' are defined in an obvious way. One checks that W' and W'' are isomorphic to U (for instance, take for basis of W' the vectors $a = (\frac{1}{2}\ \frac{1}{2})$, $b = (1, -1)$; one has $a.a = b.b = 0$, $a.b = 1$; for W'', take $(\frac{1}{2}, -\frac{1}{2})$ and $(1, 1)$). Let then $f\colon W \oplus F_1 \to W \oplus F_2$ be an isomorphism. It extends to an isomorphism of V_1 onto V_2, which carries E_1 onto E_2, thus also E_1^0 onto E_2^0 and E_1^+ onto E_2^+. Thus it carries also (E_1', E_1'') onto either (E_2', E_2'') or (E_2'', E_2'). Since E_i' and E_i'' are isomorphic to $U \oplus F_i$, one sees that $U \oplus F_1 \simeq U \oplus F_2$.

Proof of theorem 5.—We first prove that *if $E_1, E_2 \in S$ are indefinite of type II and have the same rank and same index, they are isomorphic.* By lemma 5, one has $E_1 = U \oplus F_1$, $E_2 = U \oplus F_2$; it is clear that F_1 and F_2 are of type II and same rank and same index. The modules $I_+ \oplus I_- \oplus F_1$ and $I_+ \oplus I_- \oplus F_2$ are indefinite, of type I, of same rank and index. By theorem 4, they are isomorphic. Applying lemma 6, we see then that E_1 and E_2 are isomorphic, which proves our assertion.

Theorem 5 is now clear: if E is indefinite, of type II, and if $\tau(E) \geqq 0$ we determine integers p and q by the formulas

$$q = \tfrac{1}{8}\tau(E), \qquad p = \tfrac{1}{2}(r(E) - \tau(E)).$$

By applying the above result to the modules E and $pU \oplus q\Gamma_8$ one sees that these modules are isomorphic.

Part II

Analytic Methods

Chapter VI

The Theorem on Arithmetic Progressions

The aim of this chapter is to prove the following theorem, conjectured (and used) by Legendre and proved by Dirichlet.

Theorem.—*Let a and m be relatively prime integers $\geqq 1$. There exist infinitely many prime numbers p such that $p \equiv a \pmod{m}$.*

The method we follow (which is that of Dirichlet himself) uses the properties of the *L-functions.*

§1. *Characters of finite abelian groups*

1.1. *Duality*

Let G be a finite abelian group written multiplicatively.

Definition 1.—*A character of G is a homomorphism of G into the multiplicative group $\mathbf{C}^*$ of complex numbers.*

The characters of G form a group Hom $(G, \mathbf{C}^*)$ which we denote by $\hat{G}$; it is called the *dual* of G.

Example. Suppose that G is *cyclic of order n* with generator s. If $\chi: G \to \mathbf{C}^*$ is a character of G, the element $w = \chi(s)$ satisfies the relation $w^n = 1$, i.e. is a nth root of unity. Conversely every nth root of unity w defines a character of G by means of $s^a \mapsto w^a$. Thus we see that the map $\chi \mapsto \chi(s)$ is an *isomorphism of $\hat{G}$ on the group μ_n of nth roots of unity.* In particular, $\hat{G}$ is cyclic of order n.

Proposition 1.—*Let H be a subgroup of G. Every character of H extends to a character of G.*

We use induction on the index $(G:H)$ of H in G. If $(G:H) = 1$, then $H = G$ and there is nothing to prove. Otherwise let x be an element of G not contained in H, and let n be the smallest integer > 1 such that $x^n \in H$. Let χ be a character of H, and let $t = \chi(x^n)$. Since $\mathbf{C}^*$ is a *divisible* group, one can choose an element $w \in \mathbf{C}^*$ such that $w^n = t$. Let H' be the subgroup of G generated by H and x; every element h' of H' can be written $h' = hx^a$ with $a \in \mathbf{Z}$ and $h \in H$. Put

$$\chi'(h') = \chi(h)w^a.$$

One checks that this number does not depend on the decomposition hx^a of h' and that $\chi': H' \to \mathbf{C}^*$ is a character of H' extending χ. Since we have $(G: H') < (G:H)$ the inductive hypothesis allows us to extend χ' to a character of G.

Remark. The operation of restriction defines a homomorphism

$$\rho: \hat{G} \to \hat{H}$$

and prop. 1 says that ρ is surjective. Moreover, the kernel of ρ is the set of characters of G which are trivial on H; it is thus isomorphic to the group $\widehat{(G/H)}$ dual to G/H. Hence the *exact sequence*:

$$\{1\} \to \widehat{(G/H)} \to \hat{G} \to \hat{H} \to \{1\}.$$

Proposition 2.—*The group $\hat{G}$ is a finite abelian group of the same order as G.*

One uses induction on the order n of G, the case $n = 1$ being trivial. If $n \geqq 2$, choose a nontrivial cyclic subgroup H of G. By the remark above, the order of $\hat{G}$ is the product of the orders of $\hat{H}$ and of $\widehat{(G/H)}$. But the order of H (resp. of G/H) is equal to that of its dual, because H is cyclic (resp. because G/H is of order strictly smaller than n). We conclude from this that the order of $\hat{G}$ is the product of the orders of H and of G/H, hence is equal to the order of G.

Remark. One can prove a more precise result: $\hat{G}$ is *isomorphic* (non-canonically in general) to G. This is shown by decomposing G into a product of cyclic groups.

If $x \in G$ the function $\chi \mapsto \chi(x)$ is a character of $\hat{G}$. We obtain thus a homomorphism $\varepsilon: G \to \hat{\hat{G}}$.

Proposition 3.—*The homomorphism ε is an isomorphism of G onto its bidual $\hat{\hat{G}}$.*

Since G and $\hat{\hat{G}}$ have the same order, it suffices to prove that ε is injective, i.e. that, if $x \in G$ is $\neq 1$, there exists a character χ of G such that $\chi(x) \neq 1$. Now, let H be the cyclic subgroup of G generated by x. It is clear (see the above example) that there exists a character χ of H such that $\chi(x) \neq 1$ and prop. 1 shows that χ extends to a character of G; hence the desired result.

1.2. *Orthogonality relations*

Proposition 4.—*Let $n = \mathrm{Card}(G)$ and let $\chi \in \hat{G}$. Then*

$$\sum_{x \in G} \chi(x) = \begin{cases} n & \text{if } \chi = 1 \\ 0 & \text{if } \chi \neq 1. \end{cases}$$

The first formula is obvious. To prove the second, choose $y \in G$ such that $\chi(y) \neq 1$. One has:

$$\chi(y) \sum_{x \in G} \chi(x) = \sum_{x \in G} \chi(xy) = \sum_{x \in G} \chi(x),$$

hence:

$$(\chi(y) - 1) \sum_{x \in G} \chi(x) = 0.$$

Since $\chi(y) \neq 1$, this implies $\sum_{x \in G} \chi(x) = 0$.

Corollary.—*Let $x \in G$. Then*

$$\sum_{\chi \in \hat{G}} \chi(x) = \begin{cases} n & \text{if} \quad x = 1 \\ 0 & \text{if} \quad x \neq 1. \end{cases}$$

This follows from prop. 4 applied to the dual group $\hat{G}$.

Remark.—The above results are special cases of the "orthogonality relations" of the character theory of finite groups (not necessarily abelian).

1.3. *Modular characters*

Let m be an integer $\geqq 1$. We denote by $G(m)$ the multiplicative group $(\mathbf{Z}/m\mathbf{Z})^*$ of invertible elements of the ring $\mathbf{Z}/m\mathbf{Z}$. It is an abelian group of order $\phi(m)$, where $\phi(m)$ is the Euler ϕ-function of m, cf. chap. I, n° 1.2. An element χ of the dual of $G(m)$ is called a *character modulo m*; it can be viewed as a function, defined on the set of integers prime to m, with values in $\mathbf{C}^*$, and such that $\chi(ab) = \chi(a)\chi(b)$; it is convenient to extend such a function to all of $\mathbf{Z}$ by putting $\chi(a) = 0$ if a is not prime to m.

Examples

1) $m = 4$; the group $G(4)$ has two elements, hence has a unique nontrivial character, which is $x \mapsto (-1)^{\varepsilon(x)}$, cf. chap. I, n° 3.2.

2) $m = 8$; the group $G(8)$ has four elements. It has three nontrivial characters which are

$$x \mapsto (-1)^{\varepsilon(x)}, \quad (-1)^{\omega(x)}, \quad (-1)^{\varepsilon(x)+\omega(x)}$$

cf. chap. I, n° 3.2.

3) $m = p$ with p prime $\neq 2$. The group $G(p)$ is cyclic of order $p-1$, hence has a unique character of order 2, the Legendre character $x \mapsto \left(\frac{x}{p}\right)$.

4) $m = 7$. The group $G(7)$ is cyclic of order 6, thus has two characters of order 3 which are complex conjugates. One of them is given by

$$\chi(x) = 1 \qquad \text{if } x \equiv \pm 1 \pmod 7$$

$$\chi(x) = e^{2\pi i/3} \text{ if } x \equiv \pm 2 \pmod 7$$

$$\chi(x) = e^{4\pi i/3} \text{ if } x \equiv \pm 3 \pmod 7.$$

The characters of order 2 are closely related to the Legendre characters. More precisely:

Proposition 5.—*Let a be a non-zero square-free integer* (cf. chap. IV, n° 3.2) *and let $m = 4|a|$. Then there exists a unique character χ_a modulo m such that $\chi_a(p) = \left(\frac{a}{p}\right)$ for all prime numbers p not dividing m. One has $\chi_a^2 = 1$ and $\chi_a \neq 1$ if $a \neq 1$.*

The uniqueness of χ_a is clear because all integers prime to m are products of prime numbers not dividing m; the same argument shows that $\chi_a^2 = 1$.

To prove the existence of χ_a, assume first that $a = l_1 \ldots l_k$ where the l_i are distinct prime numbers, different from 2. Then we take for χ_a the character

$$\chi_a(x) = (-1)^{\varepsilon(x)\varepsilon(a)}\left(\frac{x}{l_1}\right)\cdots\left(\frac{x}{l_k}\right).$$

If p is a prime number distinct from 2 and the l_i, the quadratic reciprocity law shows that

$$\chi_a(p) = \left(\frac{l_1}{p}\right)\cdots\left(\frac{l_k}{p}\right) = \left(\frac{a}{p}\right)$$

and χ_a has the required property. We have $\chi_a \neq 1$ if $a \neq 1$; indeed, if we choose x such that

$$\left(\frac{x}{l_1}\right) = -1 \quad \text{and} \quad x \equiv 1 \pmod{4l_2 \ldots l_k},$$

we have $\chi_a(x) = -1$.

When a is of the form $-b$ (or $2b$ or $-2b$) with $b = l_1 \ldots l_k$ as above, we take for χ_a the product of χ_b with the character $(-1)^{\varepsilon(x)}$ (or $(-1)^{\omega(x)}$ or $(-1)^{\varepsilon(x)+\omega(x)}$). A similar argument shows that $\chi_a \neq 1$.

Remark.—One can prove that, if x is an integer >0 prime to m, then

$$\chi_a(x) = \prod_{l \mid m} (a, x)_l = \prod_{(l, m) = 1} (a, x)_l$$

where $(a, x)_l$ denotes the Hilbert symbol of a and x in the field $\mathbf{Q}_l$. This formula could have been used to define χ_a.

§2. *Dirichlet series*

2.1. *Lemmas*

Lemma 1.—*Let U be an open subset of* $\mathbf{C}$ *and let f_n be a sequence of holomorphic functions on U which converges uniformly on every compact set to a function f. Then f is holomorphic in U and the derivatives f_n' of the f_n converge uniformly on all compact subsets to the derivative f' of f.*

Let us recall briefly the proof:

Let D be a closed disc contained in U and let C be its boundary oriented in the usual manner. By Cauchy formula, one has

$$f_n(z_0) = \frac{1}{2i\pi}\int_C \frac{f_n(z)}{z - z_0}\,dz$$

for all z_0 interior to D. Passing to the limit, one gets:

$$f(z_0) = \frac{1}{2i\pi}\int_C \frac{f(z)}{z - z_0}\,dz,$$

which shows that f is holomorphic in the interior of D, and the first part of the lemma follows. The second part is proved in the same way, using the formula:

$$f'(z_0) = \frac{1}{2i\pi}\int_C \frac{f(z)}{(z-z_0)^2}\,dz.$$

Lemma 2 (Abel's lemma).—*Let (a_n) and (b_n) be two sequences. Put:*

$$A_{m,p} = \sum_{n=m}^{n=p} a_n \quad \textit{and} \quad S_{m,m'} = \sum_{n=m}^{n=m'} a_n b_n.$$

Then one has:

$$S_{m,m'} = \sum_{n=m}^{n=m'-1} A_{m,n}(b_n - b_{n+1}) + A_{m,m'} b_{m'}.$$

One replaces a_n by $A_{m,n} - A_{m,n-1}$ and regroups the terms.

Lemma 3.—*Let α, β be two real numbers with $0 < \alpha < \beta$. Let $z = x+iy$ with $x, y \in \mathbf{R}$ and $x > 0$. Then:*

$$|e^{-\alpha z} - e^{-\beta z}| \leq \left|\frac{z}{x}\right| (e^{-\alpha x} - e^{-\beta x}).$$

One writes

$$e^{-\alpha z} - e^{-\beta z} = z \int_\alpha^\beta e^{-tz}\,dt,$$

hence by taking absolute values:

$$|e^{-\alpha z} - e^{-\beta z}| \leq |z| \int_\alpha^\beta e^{-tx}dt = \frac{|z|}{x}(e^{-\alpha x} - e^{-\beta x}).$$

2.2. *Dirichlet series*

Let (λ_n) be an increasing sequence of real numbers tending to $+\infty$. For the sake of simplicity, we suppose that the λ_n are ≥ 0 (this is not essential, for we can always reduce ourselves to this case by suppressing a finite number of terms of the series under consideration).

A Dirichlet series with exponents (λ_n) is a series of the form

$$\sum a_n e^{-\lambda_n z} \quad (a_n \in \mathbf{C}, z \in \mathbf{C}).$$

Examples

(a) $\lambda_n = \log n$ (ordinary Dirichlet series); such a series is written $\Sigma\, a_n/n^z$, cf. n° 2.4.

(b) $\lambda_n = n$. By setting $t = e^{-z}$, the series becomes a power series in t.

Remark.—The notion of Dirichlet series is a special case of that of the Laplace transform of a measure μ. This is the function

$$\int_0^\infty e^{-zt}\,\mu(t).$$

The case considered here is that where μ is a discrete measure. (For more details, see for instance D. WIDDER, *The Laplace Transform*, Princeton Univ. Press, 1946.)

Proposition 6.—*If the series* $f(z) = \Sigma a_n e^{-\lambda_n z}$ *converges for* $z = z_0$, *it converges uniformly in every domain of the form* $R(z-z_0) \geqq 0$, $\text{Arg}(z-z_0) \leqq \alpha$ *with* $\alpha < \pi/2$.

(Here, and in all that follows, $R(z)$ denotes the real part of the complex number z.)

After making a translation on z, we can suppose that $z_0 = 0$. The hypothesis then means that the series Σa_n is convergent. We must prove that there is uniform convergence in every domain of the form $R(z) \geqq 0$, $|z|/R(z) \leqq k$. Let $\varepsilon > 0$. Since the series Σa_n converges, there is an N such that if $m, m' \geqq N$, we have $|A_{m,m'}| \leqq \varepsilon$ (notations being those of lemma 2). Applying this lemma with $b_n = e^{-\lambda_n z}$, we obtain

$$S_{m,m'} = \sum_{m}^{m'-1} A_{m,n}(e^{-\lambda_n z} - e^{-\lambda_{n+1} z}) + A_{m,m'} e^{-\lambda_{m'} z}.$$

By putting $z = x+iy$ and applying lemma 3, we find:

$$|S_{m,m'}| \leqq \varepsilon \left(1 + \frac{|z|}{x} \sum_{m}^{m'-1} (e^{-\lambda_n x} - e^{-\lambda_{n+1} x})\right),$$

that is to say:

$$|S_{m,m'}| \leqq \varepsilon\,(1 + k(e^{-\lambda_m x} - e^{-\lambda_{m'} x}))$$

hence:

$$|S_{m,m'}| \leqq \varepsilon(1+k),$$

and the uniform convergence is clear.

Corollary 1.—*If f converges for $z = z_0$, it converges for $R(z) > R(z_0)$ and the function thus defined is holomorphic.*

This follows from prop. 6 and lemma 1.

Corollary 2.—*The set of convergence of the series f contains a maximal open half plane* (called the *half plane of convergence*).

(By abuse of language we consider $\varnothing$ and **C** as open half planes.)

If the half plane of convergence is given by $R(z) > \rho$, we say that ρ is the *abscissa of convergence* of the series considered.

(The cases $\varnothing$ and **C** correspond respectively to $\rho = +\infty$ and $\rho = -\infty$).

The half plane of convergence of the series $\Sigma |a_n| e^{-\lambda_n z}$ is called (for obvious reasons) the *half plane of absolute convergence* of f; its abscissa of convergence is denoted by ρ^+. When $\lambda_n = n$ (power series), it is well known that $\rho = \rho^+$. This is not true in general. For example the simplest L series:

$$L(z) = 1 - 1/3^z + 1/5^z - 1/7^z + \ldots$$

corresponds to $\rho = 0$ and $\rho^+ = 1$, as we will see later.

Corollary 3.—*$f(z)$ converges to $f(z_0)$ when $z \to z_0$ in the domain*

$$R(z-z_0) \geqq 0, \ |\text{Arg}\ (z-z_0)| \leqq \alpha \ \textit{with} \ \alpha < \pi/2.$$

This follows from the uniform convergence and the fact that $e^{-\lambda_n z}$ tends to $e^{-\lambda_n z_0}$.

Corollary 4.—*The function $f(z)$ can be identically zero only if all its coefficients a_n are zero.*

Let us show a_0 is zero. We multiply f by $e^{\lambda_0 z}$ and make z tend to $+\infty$ (with z real for instance). The uniform convergence shows that $e^{\lambda_0 z}f$ tends then to a_0 hence $a_0 = 0$. We proceed similarly for a_1, etc.

2.3. *Dirichlet series with positive coefficients*

Proposition 7.—*Let $f = \Sigma a_n e^{-\lambda_n z}$ be a Dirichlet series whose coefficients a_n are real $\geqq 0$. Suppose that f converges for $R(z) > \rho$, with $\rho \in \mathbf{R}$, and that the function f can be extended analytically to a function holomorphic in a neighborhood of the point $z = \rho$. Then there exists a number $\varepsilon > 0$ such that f converges for $R(z) > \rho - \varepsilon$.*

(In other terms, the domain of convergence of f is bounded by a *singularity* of f located on the real axis.)

After replacing z by $z-\rho$, we can assume that $\rho = 0$. Since f is holomorphic for $R(z) > 0$ and in a neighborhood of 0, it is holomorphic in a disc $|z-1| \leqq 1+\varepsilon$, with $\varepsilon > 0$. In particular, its Taylor series converges in this disc. By lemma 1, the pth derivative of f is given by the formula

$$f^{(p)}(z) = \sum_n a_n(-\lambda_n)^p e^{-\lambda_n z} \quad \text{for } R(z) > 0;$$

hence

$$f^{(p)}(1) = (-1)^p \sum_n \lambda_n^p a_n e^{-\lambda_n}.$$

The Taylor series in question can be written

$$f(z) = \sum_{p=0}^{\infty} \frac{1}{p!}(z-1)^p f^{(p)}(1), \qquad |z-1| \leqq 1+\varepsilon.$$

In particular for $z = -\varepsilon$, one has

$$f(-\varepsilon) = \sum_{p=0}^{\infty} \frac{1}{p!}(1+\varepsilon)^p(-1)^p f^{(p)}(1),$$

the series being convergent.

But $(-1)^p f^{(p)}(1) = \sum_n \lambda_n^p a_n e^{-\lambda_n}$ is a convergent series with positive terms.

Hence the double series with positive terms

$$f(-\varepsilon) = \sum_{p,n} a_n \frac{1}{p!}(1+\varepsilon)^p \lambda_n^p e^{-\lambda_n}$$

converges. Rearranging terms, one gets

$$f(-\varepsilon) = \sum_n a_n e^{-\lambda_n} \sum_{p=0}^{\infty} \frac{1}{p!} (1+\varepsilon)^p \lambda_n^p$$
$$= \sum_n a_n e^{-\lambda_n} e^{\lambda_n(1+\varepsilon)} = \sum_n a_n e^{\lambda_n \varepsilon},$$

which shows that the given Dirichlet series converges for $z = -\varepsilon$, thus also for $R(z) > -\varepsilon$.

2.4. *Ordinary Dirichlet series*

This is the case $\lambda_n = \log n$. The corresponding series is written

$$f(s) = \sum_{n=1}^{\infty} a_n/n^s,$$

the notation s being traditional for the variable.

Proposition 8.—*If the a_n are bounded, there is absolute convergence for $R(s) > 1$.*

This follows from the well known convergence of $\sum_{n=1}^{\infty} 1/n^\alpha$ for $\alpha > 1$.

Proposition 9.—*If the partial sums $A_{m,p} = \sum_m^p a_n$ are bounded, there is convergence (not necessarily absolute) for $R(s) > 0$.*

Assume that $|A_{m,p}| \leqq K$. By applying Abel's lemma (lemma 2), one finds

$$|S_{m,m'}| \leqq K\left(\sum_m^{m'-1} \left|\frac{1}{n^s} - \frac{1}{(n+1)^s}\right| + \left|\frac{1}{m'^s}\right|\right)$$

We can suppose that s is real (by prop. 6). This allows us to write the preceding inequality in the simpler form

$$|S_{m,m'}| \leqq K/m^s.$$

and the convergence is clear.

§3. *Zeta function and L functions*

3.1. *Eulerian products*

Definition 2.—*A function f: $\mathbf{N} \to \mathbf{C}$ is called multiplicative if $f(1) = 1$ and*

$$f(mn) = f(m)f(n)$$

whenever the integers n and m are relatively prime.

Examples.—The Euler ϕ-function (chap. I, n° 1.2) and the *Ramanujan function* (chap. VII, n° 4.5) are multiplicative.

Let f be a bounded multiplicative function.

Lemma 4.—*The Dirichlet series* $\sum_{n=1}^{\infty} f(n)/n^s$ *converges absolutely for* $R(s) > 1$ *and its sum in this domain is equal to the convergent infinite product*

$$\prod_{p \in P} (1 + f(p)p^{-s} + \dots + f(p^m)p^{-ms} + \dots).$$

(Here and in the following, P denotes the set of prime numbers.)

The absolute convergence of the series follows from the fact that f is bounded (prop. 8). Let S be a finite set of prime numbers and let $\mathbf{N}(S)$ be the set of integers $\geqq 1$ all of whose prime factors belong to S.

The following equality is immediate:

$$\sum_{n \in \mathbf{N}(S)} f(n)/n^s = \prod_{p \in S} \left(\sum_{m=0}^{\infty} f(p^m)p^{-ms} \right)$$

When S increases, the left hand side tends to $\sum_{n=1}^{\infty} f(n)/n^s$. From this, one sees that the infinite product converges and that its value is equal to $\Sigma f(n)/n^s$.

Lemma 5.—*If f is multiplicative in the strict sense* (*i.e. if* $f(nn') = f(n)f(n')$ *for all pairs* n, $n' \in \mathbf{N}$), *one has*:

$$\sum_{n=1}^{\infty} f(n)/n^s = \prod_{p \in P} \frac{1}{1 - f(p)/p^s} .$$

This follows from the above lemma and the identity $f(p^m) = f(p)^m$.

3.2. *The zeta function*

Apply the preceding section with $f = 1$. We obtain the function

$$\zeta(s) = \sum_{n=1}^{\infty} \frac{1}{n^s} = \prod_{p \in P} \frac{1}{1 - \dfrac{1}{p^s}},$$

these formulas making sense for $R(s) > 1$.

Proposition 10.—(a) *The zeta function is holomorphic and* $\neq 0$ *in the half plane* $R(s) > 1$.

(b) *One has*:

$$\zeta(s) = \frac{1}{s-1} + \phi(s),$$

where $\phi(s)$ *is holomorphic for* $R(s) > 0$.

Assertion (a) is clear. For (b), we remark that

$$\frac{1}{s-1} = \int_1^\infty t^{-s}dt = \sum_{n=1}^{\infty} \int_n^{n+1} t^{-s}dt.$$

Hence we can write

$$\zeta(s) = \frac{1}{s-1} + \sum_{n=1}^{\infty}\left(\frac{1}{n^s} - \int_n^{n+1} t^{-s}dt\right) = \frac{1}{s-1} + \sum_{n=1}^{\infty}\int_n^{n+1}(n^{-s}-t^{-s})dt.$$

Put now:

$$\phi_n(s) = \int_n^{n+1}(n^{-s}-t^{-s})dt \quad\text{and}\quad \phi(s) = \sum_{n=1}^{\infty}\phi_n(s).$$

We have to show that $\phi(s)$ is defined and holomorphic for $R(s) > 0$. But it is clear that each $\phi_n(s)$ has those properties, thus it suffices to prove that the series $\Sigma\phi_n$ converges normally on all compact sets for $R(s) > 0$. One has:

$$|\phi_n(s)| \leqq \sup_{n \leqq t \leqq n+1} |n^{-s}-t^{-s}|.$$

But the derivative of the function $n^{-s}-t^{-s}$ is equal to s/t^{s+1}. From this we get:

$$|\phi_n(s)| \leqq \frac{|s|}{n^{x+1}}, \qquad \text{with } x = R(s),$$

and the series $\Sigma\phi_n$ converges normally for $R(s) \geqq \varepsilon$, for all $\varepsilon > 0$.

Corollary 1.—*The zeta function has a simple pole for $s = 1$.*
This is clear.

Corollary 2.—*When $s \to 1$, one has $\sum_p p^{-s} \sim \log 1/(s-1)$, and $\sum_{p,k \geqq 2} 1/p^{ks}$ remains bounded.*
One has:

$$\log \zeta(s) = \sum_{p\in P, k\geqq 1}\frac{1}{k.p^{ks}} = \sum_{p\in P} 1/p^s + \psi(s),$$

with $\psi(s) = \sum_{p\in P}\sum_{k\geqq 2}(1/k.p^{ks})$. The series ψ is majorized by the series

$$\sum 1/p^{ks} = \sum 1/p^s(p^s-1) \leqq \sum 1/p(p-1) \leqq \sum_{n=2}^{\infty} 1/n(n-1) = 1.$$

This implies that ψ is bounded, and since cor. 1 shows that $\log \zeta(s) \sim \log \dfrac{1}{s-1}$, cor. 2 follows.

Remark.—Even though it is not necessary for our purpose, it should be mentioned that $\zeta(s)$ can be extended to a meromorphic function on **C** with the single pole $s = 1$. The function $\xi(s) = \pi^{-s/2}\Gamma(s/2)\zeta(s)$ is meromorphic and satisfies the *functional equation* $\xi(s) = \xi(1-s)$.

Moreover, the zeta function takes rational values on the negative integers:

$$\zeta(-2n) = 0 \qquad \text{if } n > 0$$
$$\zeta(1-2n) = (-1)^n B_n/2n \quad \text{if } n > 0,$$

where B_n denotes the nth Bernoulli number (cf. Chap. VII, n° 4.1).

One conjectures (Riemann hypothesis) that the other zeros of ζ are on the line $R(s) = \frac{1}{2}$. This has been verified numerically for a large number of them (more than three million).

3.3. *The L-functions*

Let m be an integer $\geqq 1$ and let χ be a character mod m (cf. n° 1.3). The corresponding L function is defined by the Dirichlet series

$$L(s, \chi) = \sum_{n=1}^{\infty} \chi(n)/n^s.$$

Note that, in this sum, it is only the integers n which are prime to m which give a non-zero contribution.

The case of the unit character gives nothing essentially new:

Proposition 11.—*For $\chi = 1$, one has*

$$L(s, 1) = F(s)\zeta(s) \quad \textit{with } F(s) = \prod_{p|m} (1-p^{-s}).$$

In particular $L(s, 1)$ extends analytically for $R(s) > 0$ and has a simple pole at $s = 1$.

This is clear.

Proposition 12.—*For $\chi \neq 1$ the series $L(s, \chi)$ converges (resp. converges absolutely) in the half plane $R(s) > 0$ (resp. $R(s) > 1$); one has*

$$L(s, \chi) = \prod_{p \in P} \frac{1}{1 - \dfrac{\chi(p)}{p^s}} \qquad \textit{for } R(s) > 1.$$

The assertions relative to $R(s) > 1$ follow from what has been said in n° 3.1. It remains to show the convergence of the series for $R(s) > 0$. Using prop. 9, it suffices to see that the sums

$$A_{u,v} = \sum_{u}^{v} \chi(n), \quad u \leqq v,$$

are bounded. Now, by prop. 4, we have

$$\sum_{u}^{u+m-1} \chi(n) = 0.$$

Hence it suffices to majorize the sums $A_{u,v}$ for $v-u < m$, and this is obvious: one has

$$|A_{u,v}| \leqq \phi(m).$$

The proposition follows.

Remark.—In particular $L(1, \chi)$ is *finite* when $\chi \neq 1$. The *essential point* of Dirichlet's proof consists in showing that $L(1, \chi)$ is different from zero. This is the object of the next section.

3.4. *Product of the L functions relative to the same integer m*

In this section, m is a fixed integer $\geqq 1$. If p does not divide m, we denote by $\bar{p}$ its image in $G(m) = (\mathbf{Z}/m\mathbf{Z})^*$ and by $f(p)$ the order of $\bar{p}$ in the group $G(m)$. By definition, $f(p)$ is the smallest integer $f > 1$ such that $p^f \equiv 1 \pmod{m}$. We put

$$g(p) = \phi(m)/f(p);$$

This is the order of the quotient of $G(m)$ by the subgroup $(\bar{p})$ generated by $\bar{p}$.

Lemma 6.—*If $p \nmid m$, one has the identity*

$$\prod (1-\chi(p)T) = (1-T^{f(p)})^{g(p)},$$

where the product extends over all characters χ of $G(m)$.

Let W be the set of $f(p)$-th roots of unity. One has the identity

$$\prod_{w \in W} (1-wT) = 1-T^{f(p)}.$$

Lemma 6 follows from this and the fact that for all $w \in W$ there exists $g(p)$ characters χ of $G(m)$ such that $\chi(\bar{p}) = w$.

We now define a new function $\zeta_m(s)$ by means of the formula

$$\zeta_m(s) = \prod_{\chi} L(s, \chi),$$

the product being extended over all characters χ of $G(m)$.

Proposition 13.—*One has*

$$\zeta_m(s) = \prod_{p \nmid m} \frac{1}{\left(1 - \dfrac{1}{p^{f(p)s}}\right)^{g(p)}}.$$

This is a Dirichlet series, with positive integral coefficients, converging in the half plane $R(s) > 1$.

Replacing each L function by its product expansion, and applying lemma 6 (with $T = p^{-s}$), we obtain the product expansion of $\zeta_m(s)$. This expansion shows clearly that it is a series with positive coefficients; its convergence for $R(s) > 1$ is clear.

Theorem 1.—(a) *ζ_m has a simple pole at $s = 1$.*
(b) *$L(1, \chi) \neq 0$ for all $\chi \neq 1$.*

If $L(1, \chi) \neq 0$ for all $\chi \neq 1$, the fact that $L(s, 1)$ has a simple pole at $s = 1$ shows that the same is true for ζ_m. Thus (b) $\Rightarrow$ (a). Suppose now that $L(1, \chi) = 0$ for some $\chi \neq 1$. Then the function ζ_m would be holomorphic at $s = 1$, thus also for all s such that $R(s) > 0$ (cf. prop. 11 and 12). Since it is a Dirichlet series with positive coefficients, this series would converge for all s in the same domain (prop. 7). But this is absurd. Indeed, the pth-factor of ζ_m is equal to

$$\frac{1}{(1-p^{-f(p)s})^{g(p)}} = (1+p^{-f(p)s}+p^{-2f(p)s}+\ldots)^{g(p)},$$

and dominates the series

$$1+p^{-\phi(m)s}+p^{-2\phi(m)s}+\ldots$$

It follows that ζ_m has all its coefficients greater than those of the series

$$\sum_{(n,\, m)=1} n^{-\phi(m)s}$$

which diverges for $s = \dfrac{1}{\phi(m)}$. This concludes the proof.

Remark.—The function ζ_m is equal (up to a finite number of factors) to the zeta function associated with the field of mth roots of unity. The fact that ζ_m has a simple pole at $s = 1$ can also be deduced from general results on zeta functions of algebraic number fields.

§4. *Density and Dirichlet theorem*

4.1. *Density*

Let P be the set of prime numbers. We have seen (cor. 2 to prop. 10) that, when s tends to 1 (s being real >1 to fix the ideas) one has

$$\sum_{p\in P}\frac{1}{p^s} \sim \log\frac{1}{s-1}.$$

Let A be a subset of P. One says that A has for *density* a real number k when the ratio

$$\left(\sum_{p\in A}\frac{1}{p^s}\right)\Bigg/\left(\log\frac{1}{s-1}\right)$$

tends to k when $s \to 1$. (Of course, one then has $0 \leq k \leq 1$.) The theorem on arithmetic progressions can be refined in the following way:

Theorem 2.—*Let $m \geq 1$ and let a be such that $(a, m) = 1$. Let P_a be the set of prime numbers such that $p \equiv a \pmod m$. The set P_a has density $1/\phi(m)$.*

(In other words the prime numbers are "equally distributed" between the different classes modulo m which are prime to m.)

Corollary.—*The set P_a is infinite.*

Indeed a finite set has density zero.

4.2. *Lemmas*

Let χ be a character of $G(m)$. Put

$$f_\chi(s) = \sum_{p \nmid m} \chi(p)/p^s,$$

this series being convergent for $s > 1$.

Lemma 7.—*If $\chi = 1$, then $f_\chi \sim \log \frac{1}{s-1}$ for $s \to 1$.*

Indeed, f_1 differs from the series $\Sigma 1/p^s$ by a finite number of terms only.

Lemma 8.—*If $\chi \neq 1$, f_χ remains bounded when $s \to 1$.*

We use the logarithm of the function $L(s, \chi)$. It is necessary to make a little more precise what we mean by this (due to the fact that "log" is not properly speaking a function):

$L(s, \chi)$ is defined by the product $\prod 1/(1-\chi(p)p^s)$. For $R(s) > 1$ each factor is of the form $1/(1-\alpha)$ with $|\alpha| < 1$. We define $\log \frac{1}{1-\alpha}$ as $\sum_{n=1} \alpha^n/n$ ("principal" determination of the logarithm) and we define $\log L(s, \chi)$ by the series (clearly convergent):

$$\log L(s, \chi) = \sum \log \frac{1}{1-\chi(p)p^{-s}} \quad (R(s) > 1)$$

$$= \sum_{n,\, p} \chi(p)^n/np^{ns}$$

(Equivalent definition: take the "branch" of $\log L(s, \chi)$ in $R(s) > 1$ which becomes 0 when $s \to +\infty$ on the real axis.)

We now split $\log L(s, \chi)$ into two parts:

$$\log L(s, \chi) = f_\chi(s) + F_\chi(s)$$

with

$$F_\chi(s) = \sum_{p,\, n \geqq 2} \chi(p)^n/np^{ns}.$$

Theorem 1, together with cor. 2 of prop. 10, shows that $\log L(s, \chi)$ and $F_\chi(s)$ remain bounded when $s \to 1$. Hence the same holds for $f_\chi(s)$, which proves the lemma.

4.3. *Proof of theorem* 2

We have to study the behavior of the function

$$g_a(s) = \sum_{p \in P_a} 1/p^s$$

for $s \to 1$.

Lemma 9.—*One has*

$$g_a(s) = \frac{1}{\phi(m)} \sum_\chi \chi(a)^{-1} f_\chi(s),$$

the sum being extended over all characters χ of $G(m)$.

The function $\Sigma\chi(a)^{-1}f_\chi(s)$ can be written, by replacing f_χ by its definition:

$$\sum_{p \nmid m} \Big(\sum_\chi \chi(a^{-1})\chi(p)\Big)/p^s.$$

But $\chi(a^{-1})\chi(p) = \chi(a^{-1}p)$. By the corollary to prop. 4, we have:

$$\begin{aligned} \sum_\chi \chi(a^{-1}p) &= \phi(m) \quad \text{if } a^{-1}p \equiv 1 \pmod m \\ &= 0 \qquad\quad \text{otherwise.} \end{aligned}$$

Hence we find the function $\phi(m)g_a(s)$.

Theorem 2 is now clear. Indeed, lemma 7 shows that $f_\chi(s) \sim \log \dfrac{1}{s-1}$ for $\chi = 1$, and lemma 8 shows that all other f_χ remain bounded. Using lemma 9, we then see that $g_a(s) \sim \dfrac{1}{\phi(m)} \log \dfrac{1}{s-1}$, and this means that the density of P_a is $\dfrac{1}{\phi(m)}$, q.e.d.

4.4. *An application*

Proposition 14.—*Let a be an integer which is not a square. The set of prime numbers p such that $\left(\dfrac{a}{p}\right) = 1$ has density $\frac{1}{2}$.*

We can assume that a is square-free. Let $m = 4|a|$, let χ_a be the character (mod m) defined in prop. 5 of n° 1.3 and let $H \subset G(m)$ be the kernel of χ_a in $G(m)$. If p is a prime number not divisible by m, let $\bar{p}$ be its image in $G(m)$. We have $\left(\dfrac{a}{p}\right) = 1$ if and only if $\bar{p}$ is contained in H. By th. 2 the set of prime numbers verifying this condition has for density the inverse of the index of H in $G(m)$, that is to say $\frac{1}{2}$.

Corollary.—*Let a be an integer. If the equation $X^2 - a = 0$ has a solution modulo p for almost all $p \in P$, it has a solution in $\mathbf{Z}$.*

Remark.—There are analogous results for other types of equations. For instance:

i) let $f(x) = a_0X^n + \ldots + a_n$ be a polynomial of degree n with integer coefficients, which is irreducible over $\mathbf{Q}$. Let K be the field generated by the roots of f (in an algebraically closed extension of $\mathbf{Q}$) and let $N = [K\colon \mathbf{Q}]$. One has $N \geqq n$. Let P_f be the set of prime numbers p such that f "decomposes completely modulo p", i.e. such that all the roots of f (mod p) belong to $\mathbf{F}_p$. One can prove that P_f has *density* $\dfrac{1}{N}$. (The method is analogous to that of the Dirichlet theorem—one uses the fact that the zeta function of the field K has a simple pole at $s = 1$.) One can also give the density of the set P'_f of p such that the reduction of $f \pmod p$ has at least one root in $\mathbf{F}_p$; it is a number of the form q/N with $1 \leqq q < N$ (setting aside the trivial case where $n = 1$).

ii) More generally, let $\{f_\alpha(x_1, \ldots, x_n)\}$ be a family of polynomials with integer coefficients and let Q be the set of $p \in P$ such that the reductions of $f_\alpha \pmod p$ have a common zero in $(\mathbf{F}_p)^n$. It can be proved (see J. Ax, *Ann. of Maths.*, 85, 1967, pp. 161–183) that Q has a density; moreover this density is a rational number and is zero only if Q is finite.

4.5. *Natural density*

The density used in this paragraph is the "analytic density" (or "Dirichlet density"). Despite its apparent complexity, it is very convenient.

There is another notion, that of "natural density": a subset A of P has natural density k if the ratio

$$\frac{\text{number of elements of } A \text{ which are } \leqq n}{\text{number of elements of } P \text{ which are } \leqq n}$$

tends to k when $n \to \infty$.

One can prove that, if A has natural density k, the analytic density of A exists and is equal to k. On the other hand, there exist sets having an analytic density but no natural density. It is the case, for example, of the set P^1 of prime numbers whose first digit (in the decimal system, say) is equal to 1. One sees easily, using the prime number theorem, that P^1 does not have a natural density and on the other hand BOMBIERI has shown me a proof that the analytic density of P^1 exists (it is equal to $\log_{10} 2 = 0.301029995\ldots$).

However, this "pathology" does not occur for the sets of prime numbers considered above: *the set of $p \in P$ such that $p \equiv a \pmod m$ has a natural density* (equal to $1/\phi(m)$, if a is prime to m); the same holds for the sets denoted P_f, P'_f, and Q in the preceding section. For a proof (and an estimate of the "error term") see K. PRACHAR, *Primzahlverteilung*, V, §7.

Chapter VII

Modular Forms

§1. The modular group

1.1. *Definitions*

Let H denote the upper half plane of $\mathbf{C}$, i.e. the set of complex numbers z whose imaginary part $Im(z)$ is >0.

Let $\mathbf{SL}_2(\mathbf{R})$ be the group of matrices $\begin{pmatrix} a & b \\ c & d \end{pmatrix}$, with real coefficients, such that $ad-bc = 1$. We make $\mathbf{SL}_2(\mathbf{R})$ act on $\tilde{\mathbf{C}} = \mathbf{C} \cup \{\infty\}$ in the following way:

if $g = \begin{pmatrix} a & b \\ c & d \end{pmatrix}$ is an element of $\mathbf{SL}_2(\mathbf{R})$, and if $z \in \tilde{\mathbf{C}}$, we put

$$gz = \frac{az+b}{cz+d}.$$

One checks easily the formula

$$Im(gz) = \frac{Im(z)}{|cz+d|^2}. \tag{1}$$

This shows that H is *stable* under the action of $\mathbf{SL}_2(\mathbf{R})$. Note that the element $-1 = \begin{pmatrix} -1 & 0 \\ 0 & -1 \end{pmatrix}$ of $\mathbf{SL}_2(\mathbf{R})$ acts trivially on H. We can then consider that it is the group $\mathbf{PSL}_2(\mathbf{R}) = \mathbf{SL}_2(\mathbf{R})/\{\pm 1\}$ which operates (and this group acts *faithfully*—one can even show that it is the group of all analytic automorphisms of H).

Let $\mathbf{SL}_2(\mathbf{Z})$ be the subgroup of $\mathbf{SL}_2(\mathbf{R})$ consisting of the matrices with coefficients in $\mathbf{Z}$. It is a discrete subgroup of $\mathbf{SL}_2(\mathbf{R})$.

Definition 1.—*The group $G = \mathbf{SL}_2(\mathbf{Z})/\{\pm 1\}$ is called the modular group; it is the image of $\mathbf{SL}_2(\mathbf{Z})$ in $\mathbf{PSL}_2(\mathbf{R})$.*

If $g = \begin{pmatrix} a & b \\ c & d \end{pmatrix}$ is an element of $\mathbf{SL}_2(\mathbf{Z})$, we often use the same symbol to denote its image in the modular group G.

1.2. *Fundamental domain of the modular group*

Let S and T be the elements of G defined respectively by $\begin{pmatrix} 0 & -1 \\ 1 & 0 \end{pmatrix}$ and $\begin{pmatrix} 1 & 1 \\ 0 & 1 \end{pmatrix}$. One has:

$$Sz = -1/z, \qquad Tz = z+1$$
$$S^2 = 1, \qquad (ST)^3 = 1$$

On the other hand, let D be the subset of H formed of all points z such that $|z| \geqq 1$ and $|Re(z)| \leqq 1/2$. The figure below represents the transforms of D by the elements:

$\{1, T, TS, ST^{-1}S, S, ST, STS, T^{-1}S, T^{-1}\}$ of the group G.

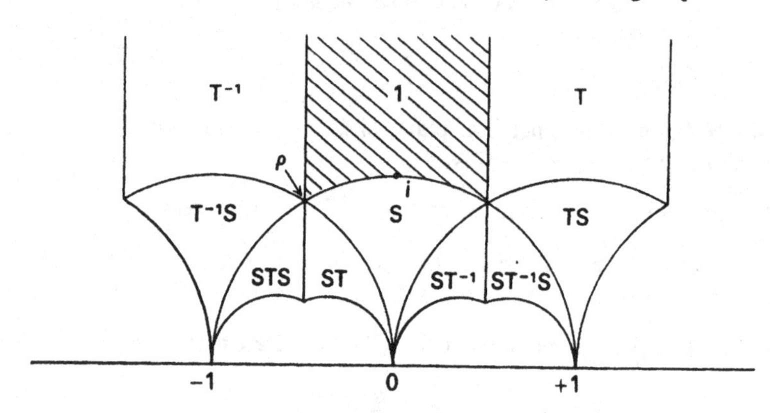

Fig. 1

We will show that D is a *fundamental domain* for the action of G on the half plane H. More precisely:

Theorem 1.—(1) *For every $z \in H$, there exists $g \in G$ such that $gz \in D$.*

(2) *Suppose that two distinct points z, z' of D are congruent modulo G. Then, $R(z) = \pm\frac{1}{2}$ and $z = z' \pm 1$, or $|z| = 1$ and $z' = -1/z$.*

(3) *Let $z \in D$ and let $I(z) = \{g | g \in G,\ gz = z\}$ the stabilizer of z in G. One has $I(z) = \{1\}$ except in the following three cases:*

$z = i$, in which case $I(z)$ is the group of order 2 generated by S;

$z = \rho = e^{2\pi i/3}$, in which case $I(z)$ is the group of order 3 generated by ST;

$z = -\bar{\rho} = e^{\pi i/3}$, in which case $I(z)$ is the group of order 3 generated by TS.

Assertions (1) and (2) imply:

Corollary.—*The canonical map $D \to H/G$ is surjective and its restriction to the interior of D is injective.*

Theorem 2.—*The group G is generated by S and T.*

Proof of theorems 1 *and* 2.—*Let G'* be the subgroup of G generated by S and T, and let $z \in H$. We are going to show that *there exists $g' \in G'$ such that $g'z \in D$*, and this will prove assertion (1) of theorem 1. If $g = \begin{pmatrix} a & b \\ c & d \end{pmatrix}$ is an element of G', then

$$(1) \qquad Im(gz) = \frac{Im(z)}{|cz+d|^2}.$$

Since c and d are integers, the numbers of pairs (c, d) such that $|cz+d|$ is less than a given number is *finite*. This shows that there exists $g \in G'$ such that $Im(gz)$ is maximum. Choose now an integer n such that $T^n gz$ has real part between $-\frac{1}{2}$ and $+\frac{1}{2}$. The element $z' = T^n gz$ *belongs to* D; indeed, it suffices to see that $|z'| \geqq 1$, but if $|z'| < 1$, the element $-1/z'$ would have an imaginary part strictly larger than $Im(z')$, which is impossible. Thus the element $g' = T^n g$ has the desired property.

We now prove assertions (2) and (3) of theorem 1. Let $z \in D$ and let $g = \begin{pmatrix} a & b \\ c & d \end{pmatrix} \in G$ such that $gz \in D$. Being free to replace (z, g) by (gz, g^{-1}), we may suppose that $Im(gz) \geqq Im(z)$, i.e. that $|cz+d|$ is $\leqq 1$. This is clearly impossible if $|c| \geqq 2$, leaving then the cases $c = 0, 1, -1$. If $c = 0$, we have $d = \pm 1$ and g is the translation by $\pm b$. Since $R(z)$ and $R(gz)$ are both between $-\frac{1}{2}$ and $\frac{1}{2}$, this implies either $b = 0$ and $g = 1$ or $b = \pm 1$ in which case one of the numbers $R(z)$ and $R(gz)$ must be equal to $-\frac{1}{2}$ and the other to $\frac{1}{2}$. If $c = 1$, the fact that $|z+d|$ is $\leqq 1$ implies $d = 0$ except if $z = \rho$ (resp. $-\bar{\rho}$) in which case we can have $d = 0, 1$ (resp, $d = 0, -1$). The case $d = 0$ gives $|z| \leq 1$ hence $|z| = 1$; on the other hand, $ad - bc = 1$ implies $b = -1$, hence $gz = a - 1/z$ and the first part of the discussion proves that $a = 0$ except if $R(z) = \pm\frac{1}{2}$, i.e. if $z = \rho$ or $-\bar{\rho}$ in which case we have $a = 0, -1$ or $a = 0, 1$. The case $z = \rho$, $d = 1$ gives $a - b = 1$ and $g\rho = a - 1/(1+\rho) = a + \rho$, hence $a = 0, 1$; we argue similarly when $z = -\bar{\rho}$, $d = -1$. Finally the case $c = -1$ leads to the case $c = 1$ by changing the signs of a, b, c, d (which does not change g, viewed as an element of G). This completes the verification of assertions (2) and (3).

It remains to prove that $G' = G$. Let g be an element of G. Choose a point z_0 *interior* to D (for example $z_0 = 2i$), and let $z = gz_0$. We have seen above that there exists $g' \in G'$ such that $g'z \in D$. The points z_0 and $g'z = g'gz_0$ of D are congruent modulo G, and one of them is interior to D. By (2) and (3), it follows that these points coincide and that $g'g = 1$. Hence we have $g \in G'$, which completes the proof.

Remark.—One can show that $\langle S, T; S^2, (ST)^3 \rangle$ is a *presentation* of G, or, equivalently, that G is the *free product of the cyclic group of order* 2 *generated by S and the cyclic group of order* 3 *generated by ST.*

§2. *Modular functions*

2.1. *Definitions*

Definition 2.—*Let k be an integer. We say a function f is weakly modular of weight* $2k$[1] *if f is meromorphic on the half plane H and verifies the relation*

$$f(z) = (cz+d)^{-2k} f\left(\frac{az+b}{cz+d}\right) \quad \textit{for all } \begin{pmatrix} a & b \\ c & d \end{pmatrix} \in \mathbf{SL}_2(\mathbf{Z}). \tag{2}$$

[1] Some authors say that f is "of weight $-2k$", others that f is "of weight k".

Let g be the image in G of $\begin{pmatrix} a & b \\ c & d \end{pmatrix}$. We have $d(gz)/dz = (cz+d)^{-2}$. The relation (2) can then be written:

$$\frac{f(gz)}{f(z)} = \left(\frac{d(gz)}{dz}\right)^{-k}$$

or

(3) $$f(gz)d(gz)^k = f(z)dz^k.$$

It means that the "differential form of weight k" $f(z)dz^k$ is *invariant* under G. Since G is generated by the elements S and T (see th. 2), it suffices to check the invariance by S and by T. This gives:

Proposition 1.—*Let f be meromorphic on H. The function f is a weakly modular function of weight $2k$ if and only if it satisfies the two relations:*

(4) $$f(z+1) = f(z)$$

(5) $$f(-1/z) = z^{2k}f(z).$$

Suppose the relation (4) is verified. We can then express f as a function of $q = e^{2\pi iz}$, function which we will denote by $\tilde{f}$; it is meromorphic in the disk $|q| < 1$ with the origin removed. If $\tilde{f}$ extends to a meromorphic (resp. holomorphic) function at the origin, we say, by abuse of language, that f is *meromorphic* (resp. *holomorphic*) *at infinity*. This means that $\tilde{f}$ admits a Laurent expansion in a neighborhood of the origin

$$\tilde{f}(q) = \sum_{-\infty}^{+\infty} a_n q^n$$

where the a_n are zero for n small enough (resp. for $n < 0$).

Definition 3.—*A weakly modular function is called modular if it is meromorphic at infinity.*

When f is holomorphic at infinity, we set $f(\infty) = \tilde{f}(0)$. This is the *value* of f at infinity.

Definition 4.—*A modular function which is holomorphic everywhere (including infinity) is called a modular form; if such a function is zero at infinity, it is called a cusp form* ("*Spitzenform*" in German—"*forme parabolique*" in French).

A modular form of weight $2k$ is thus given by a series

(6) $$f(z) = \sum_{n=0}^{\infty} a_n q^n = \sum_{n=0}^{\infty} a_n e^{2\pi inz}$$

which converges for $|q| < 1$ (i.e. for $Im(z) > 0$), and which verifies the identity

(5) $$f(-1/z) = z^{2k}f(z).$$

It is a cusp form if $a_0 = 0$.

Examples

1) If f and f' are modular forms of weight $2k$ and $2k'$, the product ff' is a modular form of weight $2k+2k'$.

2) We will see later that the function

$$q \prod_{n=1}^{\infty} (1-q^n)^{24} = q-24q^2+252q^3-1472q^4+\ldots$$

is a cusp form of weight 12.

2.2. *Lattice functions and modular functions*

We recall first what is a *lattice* in a real vector space V of finite dimension. It is a subgroup Γ of V verifying one of the following equivalent conditions:

i) Γ is discrete and V/Γ is compact;
ii) Γ is discrete and generates the $\mathbf{R}$-vector space V;
iii) There exists an $\mathbf{R}$-basis $(e_1, \ldots, e_n)$ of V which is a $\mathbf{Z}$-basis of Γ (i.e. $\Gamma = \mathbf{Z}e_1 \oplus \ldots \oplus \mathbf{Z}e_n$).

Let $\mathcal{R}$ be the *set of lattices* of $\mathbf{C}$ considered as an $\mathbf{R}$-vector space. Let M be the set of pairs (ω_1, ω_2) of elements of $\mathbf{C}^*$ such that $Im(\omega_1/\omega_2) > 0$; to such a pair we associate the lattice

$$\Gamma(\omega_1, \omega_2) = \mathbf{Z}\omega_1 \oplus \mathbf{Z}\omega_2$$

with basis $\{\omega_1, \omega_2\}$. We thus obtain a map $M \to \mathcal{R}$ which is clearly *surjective*.

Let $g = \begin{pmatrix} a & b \\ c & d \end{pmatrix} \in \mathbf{SL}_2(\mathbf{Z})$ and let $(\omega_1, \omega_2) \in M$. We put

$$\omega_1' = a\omega_1 + b\omega_2 \quad \text{and} \quad \omega_2' = c\omega_1 + d\omega_2.$$

It is clear that $\{\omega_1', \omega_2'\}$ is a basis of $\Gamma(\omega_1, \omega_2)$. Moreover, if we set $z = \omega_1/\omega_2$ and $z' = \omega_1'/\omega_2'$, we have

$$z' = \frac{az+b}{cz+d} = gz.$$

This shows that $Im(z') > 0$, hence that (ω_1', ω_2') belongs to M.

Proposition 2.—*For two elements of M to define the same lattice it is necessary and sufficient that they are congruent modulo* $\mathbf{SL}_2(\mathbf{Z})$.

We just saw that the condition is sufficient. Conversely, if (ω_1, ω_2) and (ω_1', ω_2') are two elements of M which define the same lattice, there exists an integer matrix $g = \begin{pmatrix} a & b \\ c & d \end{pmatrix}$ of determinant ± 1 which transforms the first basis into the second. If $\det(g)$ was <0, the sign of $Im(\omega_1'/\omega_2')$ would be *the opposite* of $Im(\omega_1/\omega_2)$ as one sees by an immediate computation. The two signs being the same, we have necessarily $\det(g) = 1$ which proves the proposition.

Hence we can identify the *set $\mathcal{R}$ of lattices of* $\mathbf{C}$ *with the quotient of M by the action of* $\mathbf{SL}_2(\mathbf{Z})$.

Make now $\mathbf{C}^*$ act on $\mathscr{R}$ (resp. on M) by:

$$\Gamma \mapsto \lambda\Gamma \quad (\text{resp. } (\omega_1, \omega_2) \mapsto (\lambda\omega_1, \lambda\omega_2)), \quad \lambda \in \mathbf{C}^*.$$

The quotient $M/\mathbf{C}^*$ is identified with H by $(\omega_1, \omega_2) \mapsto z = \omega_1/\omega_2$, and this identification transforms the action of $\mathbf{SL}_2(\mathbf{Z})$ on M into that of $G = \mathbf{SL}_2(\mathbf{Z})/\{\pm 1\}$ on H (cf. n° 1.1). Hence:

Proposition 3.—*The map* $(\omega_1, \omega_2) \mapsto \omega_1/\omega_2$ *gives by passing to the quotient, a bijection of* $\mathscr{R}/\mathbf{C}^*$ *onto* H/G. (Thus, an element of H/G can be identified with a lattice of $\mathbf{C}$ defined up to a homothety.)

Remark.—Let us associate to a lattice Γ of $\mathbf{C}$ the *elliptic curve* $E_\Gamma = \mathbf{C}/\Gamma$. It is easy to see that two lattices Γ and Γ' define isomorphic elliptic curves if and only if they are homothetic. This gives a third description of $H/G = \mathscr{R}/\mathbf{C}^*$: it is the set of *isomorphism classes of elliptic curves.*

Let us pass now to *modular functions.* Let F be a function on $\mathscr{R}$, with complex values, and let $k \in \mathbf{Z}$. We say that F is of *weight* $2k$ if

$$F(\lambda\Gamma) = \lambda^{-2k}F(\Gamma) \tag{7}$$

for all lattices Γ and all $\lambda \in \mathbf{C}^*$.

Let F be such a function. If $(\omega_1, \omega_2) \in M$, we denote by $F(\omega_1, \omega_2)$ the value of F on the lattice $\Gamma(\omega_1, \omega_2)$. The formula (7) translates to:

$$F(\lambda\omega_1, \lambda\omega_2) = \lambda^{-2k}F(\omega_1, \omega_2). \tag{8}$$

Moreover, $F(\omega_1, \omega_2)$ is invariant by the action of $\mathbf{SL}_2(\mathbf{Z})$ on M.

Formula (8) shows that the product $\omega_2^{2k}F(\omega_1, \omega_2)$ depends only on $z = \omega_1/\omega_2$. There exists then a function f on H such that

$$F(\omega_1, \omega_2) = \omega_2^{-2k}f(\omega_1/\omega_2). \tag{9}$$

Writing that F is invariant by $\mathbf{SL}_2(\mathbf{Z})$, we see that f satisfies the identity:

$$f(z) = (cz+d)^{-2k}f\left(\frac{az+b}{cz+d}\right) \quad \text{for all} \begin{pmatrix} a & b \\ c & d \end{pmatrix} \in \mathbf{SL}_2(\mathbf{Z}). \tag{2}$$

Conversely, if f verifies (2), formula (9) associates to it a function F on $\mathscr{R}$ which is of weight $2k$. We can thus identify *modular functions of weight* $2k$ with some *lattice functions of weight* $2k$.

2.3. *Examples of modular functions; Eisenstein series*

Lemma 1.—*Let* Γ *be a lattice in* $\mathbf{C}$. *The series* $\sum'_{\gamma\in\Gamma} 1/|\gamma|^\sigma$ *is convergent for* $\sigma > 2$.

(The symbol Σ' signifies that the summation runs over the nonzero elements of Γ.)

We can proceed as with the series $\Sigma 1/n^\alpha$, i.e. majorize the series under consideration by a multiple of the double integral $\displaystyle\iint \frac{dxdy}{(x^2+y^2)^{\sigma/2}}$ extended over the plane deprived of a disk with center 0. The double integral is easily computed using "polar coordinates". Another method, essentially equivalent, consists in remarking that the number of elements of Γ such that $|\gamma|$ is between two consecutive integers n and $n+1$ is $O(n)$; the convergence of the series is thus reduced to that of the series $\Sigma 1/n^{\sigma-1}$.

Now let k be an integer >1. If Γ is a lattice of $\mathbf{C}$, put

$$G_k(\Gamma) = \sum_{\gamma\in\Gamma}' 1/\gamma^{2k}. \tag{10}$$

This series converges absolutely, thanks to lemma 1. It is clear that G_k is of *weight* $2k$. It is called the *Eisenstein series* of index k (or index $2k$ following other authors). As in the preceding section, we can view G_k as a function on M, given by:

$$G_k(\omega_1, \omega_2) = \sum_{m,n}' \frac{1}{(m\omega_1+n\omega_2)^{2k}}. \tag{11}$$

Here again the symbol Σ' means that the summation runs over all pairs of integers (m, n) *distinct from* $(0, 0)$. The function on H corresponding to G_k (by the procedure given in the preceding section) is again denoted by G_k. By formulas (9) and (11), we have

$$G_k(z) = \sum_{m,n}' \frac{1}{(mz+n)^{2k}}. \tag{12}$$

Proposition 4.—*Let k be an integer >1. The Eisenstein series $G_k(z)$ is a modular form of weight $2k$. We have $G_k(\infty) = 2\zeta(2k)$ where ζ denotes the Riemann zeta function.*

The above arguments show that $G_k(z)$ is *weakly modular* of weight $2k$. We have to show that G_k is everywhere holomorphic (including infinity). First suppose that z is contained in the fundamental domain D (cf. n° 1.2). Then

$$\begin{aligned}|mz+n|^2 &= m^2 z\bar{z}+2mnR(z)+n^2\\ &\geqq m^2-mn+n^2 = |m\rho-n|^2.\end{aligned}$$

By lemma 1, the series $\Sigma' 1/|m\rho-n|^{2k}$ is convergent. This shows that the series $G_k(z)$ *converges normally in* D, thus also (applying the result to $G_k(g^{-1}z)$ with $g\in G$) in each of the transforms gD of D by G. Since these cover H (th. 1), we see that G_k is holomorphic in H. It remains to see that G_k is holomorphic at infinity (and to find the value at this point). This amounts to proving that G_k has a limit for $Im(z)\to\infty$. But one may suppose that z remains in the fundamental domain D; in view of the uniform convergence in D, we can make the passage to the limit term by term. The terms $1/(mz+n)^{2k}$ relative to $m \neq 0$ give 0; the others give $1/n^{2k}$. Thus

$$\lim. G_k(z) = \sum' 1/n^{2k} = 2\sum_{n=1}^{\infty} 1/n^{2k} = 2\zeta(2k) \qquad \text{q.e.d.}$$

Remark.—We give in n° 4.2 below the expansion of G_k as a power series in $q = e^{2\pi iz}$.

Examples.—The Eisenstein series of lowest weights are G_2 and G_3, which are of weight 4 and 6. It is convenient (because of the theory of elliptic curves) to replace these by multiples:

$$g_2 = 60G_2, \qquad g_3 = 140G_3. \tag{13}$$

We have $g_2(\infty) = 120\zeta(4)$ and $g_3(\infty) = 280\zeta(6)$. Using the known values of $\zeta(4)$ and $\zeta(6)$ (see for example n° 4.1 below), one finds:

$$g_2(\infty) = \frac{4}{3}\pi^4, \qquad g_3(\infty) = \frac{8}{27}\pi^6. \tag{14}$$

If we put

$$\Delta = g_2^3 - 27g_3^2, \tag{15}$$

we have $\Delta(\infty) = 0$; that is to say, Δ *is a cusp form of weight* 12.

Relation with elliptic curves

Let Γ be a lattice of $\mathbf{C}$ and let

$$\wp_\Gamma(u) = \frac{1}{u^2} + \sum_{\gamma\in\Gamma}' \left(\frac{1}{(u-\gamma)^2} - \frac{1}{\gamma^2}\right) \tag{16}$$

be the corresponding Weierstrass function[1]. The $G_k(\Gamma)$ occur into the Laurent expansion of $\wp_\Gamma$:

$$\wp_\Gamma(u) = \frac{1}{u^2} + \sum_{k=2}^{\infty} (2k-1)G_k(\Gamma)u^{2k-2}. \tag{17}$$

If we put $x = \wp_\Gamma(u)$, $y = \wp'_\Gamma(u)$, we have

$$y^2 = 4x^3 - g_2x - g_3, \tag{18}$$

with $g_2 = 60G_2(\Gamma)$, $g_3 = 140G_3(\Gamma)$ as above. Up to a numerical factor, $\Delta = g_2^3 - 27g_3^2$ is equal to the *discriminant* of the polynomial $4x^3 - g_2x - g_3$.

One proves that the cubic defined by the equation (18) in the projective plane is isomorphic to the elliptic curve $\mathbf{C}/\Gamma$. In particular, it is a *nonsingular curve*, and this shows that Δ is $\neq 0$.

§3. *The space of modular forms*

3.1. *The zeros and poles of a modular function*

Let f be a meromorphic function on H, not identically zero, and let p be a point of H. The integer n such that $f/(z-p)^n$ is holomorphic and non-zero at p is called the *order of f at p* and is denoted by $v_p(f)$.

---

[1] See for example H. CARTAN, *Théorie élémentaire des fonctions analytiques d'une ou plusieurs variables complexes*, chap. V, §2, n° 5. (English translation: Addison-Wesley Co.)

When f is a *modular function* of weight $2k$, the identity

$$f(z) = (cz+d)^{-2k} f\left(\frac{az+b}{cz+d}\right)$$

shows that $v_p(f) = v_{g(p)}(f)$ if $g \in G$. In other terms, $v_p(f)$ depends only on the image of p in H/G. Moreover one can define $v_\infty(f)$ as the order for $q = 0$ of the function $\tilde{f}(q)$ associated to f (cf. n° 2.1).

Finally, we will denote by e_p the order of the stabilizer of the point p; we have $e_p = 2$ (resp. $e_p = 3$) if p is congruent modulo G to i (resp. to ρ) and $e_p = 1$ otherwise, cf. th. 1.

Theorem 3.—*Let f be a modular function of weight $2k$, not identically zero. One has:*

$$v_\infty(f) + \sum_{p \in H/G} \frac{1}{e_p} v_p(f) = \frac{k}{6}. \tag{19}$$

[We can also write this formula in the form

$$v_\infty(f) + \frac{1}{2} v_i(f) + \frac{1}{3} v_\rho(f) + \sum_{p \in H/G}^{*} v_p(f) = \frac{k}{6} \tag{20}$$

where the symbol Σ^* means a summation over the points of H/G distinct from the classes of i and ρ.]

Observe first that the sum written in th. 3 makes sense, i.e. that f *has only a finite number of zeros and poles* modulo G. Indeed, since $\tilde{f}$ is meromorphic, there exists $r > 0$ such that $\tilde{f}$ has no zero nor pole for $0 < |q| < r$; this means that f has no zero nor pole for $Im(z) > \frac{1}{2\pi} \log(1/r)$. Now, the part D_r of the fundamental domain D defined by the inequality $Im(z) \leqq \frac{1}{2\pi}\log(1/r)$ is *compact*; since f is meromorphic in H, it has only a finite number of zeros and of poles in D_r, hence our assertion.

To prove theorem 3, we will integrate $\frac{1}{2i\pi}\frac{df}{f}$ on the boundary of D. More precisely:

1) Suppose that f has no zero nor pole on the boundary of D except possibly i, ρ, and $-\bar{\rho}$. There exists a contour $\mathscr{C}$ as represented in Fig. 2 whose interior contains a representative of each zero or pole of f not congruent to i or ρ. By the residue theorem we have

$$\frac{1}{2\pi i}\int_{\mathscr{C}} \frac{df}{f} = \sum_{p \in H/G}^{*} v_p(f)$$

On the other hand:

a) The change of variables $q = e^{2\pi i z}$ transforms the arc EA into a circle ω centered at $q = 0$, with negative orientation, and not enclosing any zero or pole of $\tilde{f}$ except possibly 0. Hence

$$\frac{1}{2i\pi}\int_E^A \frac{df}{f} = \frac{1}{2i\pi}\int_{\omega} \frac{df}{f} = -v_\infty(f).$$

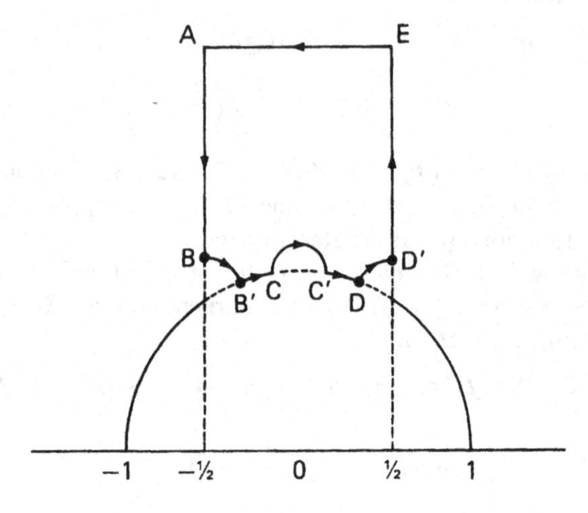

Fig. 2

b) The integral of $\frac{1}{2i\pi}\frac{df}{f}$ on the circle which contains the arc BB', oriented negatively, has the value $-v_\rho(f)$. When the radius of this circle tends to 0, the angle $\widehat{B_\rho B'}$ tends to $\frac{2\pi}{6}$. Hence:

$$\frac{1}{2i\pi}\int_B^{B'}\frac{df}{f} \to -\frac{1}{6}v_\rho(f).$$

Similarly when the radii of the arcs CC' and DD' tend to 0:

$$\frac{1}{2i\pi}\int_C^{C'}\frac{df}{f} \to -\frac{1}{2}v_i(f)$$

$$\frac{1}{2i\pi}\int_D^{D'}\frac{df}{f} \to -\frac{1}{6}v_\rho(f).$$

c) T transforms the arc AB into the arc ED'; since $f(Tz) = f(z)$, we get:

$$\frac{1}{2i\pi}\int_A^B\frac{df}{f} + \frac{1}{2i\pi}\int_{D'}^E\frac{df}{f} = 0.$$

d) S transforms the arc $B'C$ onto the arc DC'; since $f(Sz) = z^{2k}f(z)$, we get:

$$\frac{df(Sz)}{f(Sz)} = 2k\frac{dz}{z} + \frac{df(z)}{f(z)},$$

hence:

$$\frac{1}{2i\pi}\int_{B'}^{C}\frac{df}{f}+\frac{1}{2i\pi}\int_{C'}^{D}\frac{df}{f}=\frac{1}{2i\pi}\int_{B'}^{C}\left(\frac{df(z)}{f(z)}-\frac{df(Sz)}{f(Sz)}\right)$$

$$=\frac{1}{2i\pi}\int_{B'}^{C}\left(-2k\frac{dz}{z}\right)$$

$$\to -2k\left(-\frac{1}{12}\right)=\frac{k}{6}$$

when the radii of the arcs BB', CC', DD', tend to 0.

Writing now that the two expressions we get for $\frac{1}{2i\pi}\int_{\mathscr{C}}\frac{df}{f}$ are equal, and passing to the limit, we find formula (20).

2) Suppose that f has a zero or a pole λ on the half line

$$\left\{z \,|\, Re(z) = -\frac{1}{2}, Im(z) > \frac{\sqrt{3}}{2}\right\}.$$

We repeat the above proof with a contour modified in a neighborhood of λ and of $T\lambda$ as in Fig. 3. (The arc circling around $T\lambda$ is the transform by T of the arc circling around λ.)

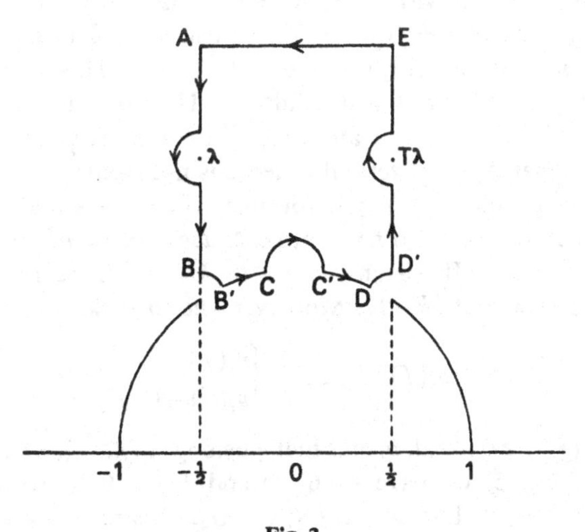

Fig. 3

We proceed in an analogous way if f has several zeros or poles on the boundary of D.

Remark.—This somewhat laborious proof could have been avoided if one had defined a complex analytic structure on the compactification of H/G

(see for instance *Seminar on Complex Multiplication*, Lecture Notes on Math., n° 21, lecture II).

3.2. *The algebra of modular forms*

If k is an integer, we denote by M_k (resp. M_k^0) the **C**-vector space of modular forms of weight $2k$ (resp. of cusp forms of weight $2k$) cf. n° 2.1, def. 4. By definition, M_k^0 is the kernel of the linear form $f \mapsto f(\infty)$ on M_k. Thus we have $\dim M_k/M_k^0 \leqq 1$. Moreover, for $k \geqq 2$, the Eisenstein series G_k is an element of M_k such that $G_k(\infty) \neq 0$, cf. n° 2.3, prop. 4. Hence we have

$$M_k = M_k^0 \oplus \mathbf{C}.G_k \quad (\text{for } k \geqq 2).$$

Finally recall that one denotes by Δ the element $g_2^3 - 27g_3^2$ of M_6^0 where $g_2 = 60G_2$ and $g_3 = 140G_3$.

Theorem 4.—(i) *We have* $M_k = 0$ *for* $k < 0$ *and* $k = 1$.

(ii) *For* $k = 0, 2, 3, 4, 5$, M_k *is a vector space of dimension* 1 *with basis* 1, G_2, G_3, G_4, G_5; *we have* $M_k^0 = 0$.

(iii) *Multiplication by* Δ *defines an isomorphism of* M_{k-6} *onto* M_k^0.

Let f be a nonzero element of M_k. All the terms on the left side of the formula

$$v_\infty(f) + \frac{1}{2}v_i(f) + \frac{1}{3}v_\rho(f) + \sum_{p \in H/G}^{*} v_p(f) = \frac{k}{6} \tag{20}$$

are ≥ 0. Thus we have $k \geqq 0$ and also $k \neq 1$, since $\frac{1}{6}$ cannot be written in the form $n + n'/2 + n''/3$ with $n, n', n'' \geqq 0$. This proves (i).

Now apply (20) to $f = G_k$, $k = 2$. We can write $\frac{2}{6}$ in the form $n + n'/2 + n''/3$, $n, n', n'' \geqq 0$ only for $n = 0$, $n' = 0$, $n'' = 1$. This shows that $v_\rho(G_2) = 1$ and $v_p(G_2) = 0$ for $p \neq \rho$ (modulo G). The same argument applies to G_3 and proves that $v_i(G_3) = 1$ and that all the others $v_p(G_3)$ are zero. This already shows that Δ is not zero at i, hence is not identically zero. Since the weight of Δ is 12 and $v_\infty(\Delta) \geqq 1$, formula (20) implies that $v_p(\Delta) = 0$ for $p \neq \infty$ and $v_\infty(\Delta) = 1$. In other words, Δ *does not vanish on* H *and has a simple zero at infinity*. If f is an element of M_k^0 and if we set $g = f/\Delta$, it is clear that g is of weight $2k - 12$. Moreover, the formula

$$v_p(g) = v_p(f) - v_p(\Delta) = \begin{cases} v_p(f) & \text{if } p \neq \infty \\ v_p(f) - 1 & \text{if } p = \infty \end{cases}$$

shows that $v_p(g)$ is $\geqq 0$ for all p, thus that g belongs to M_{k-6}, which proves (iii).

Finally, if $k \leqq 5$, we have $k - 6 < 0$ and $M_k^0 = 0$ by (i) and (iii); this shows that $\dim M_k \leqq 1$. Since $1, G_2, G_3, G_4, G_5$ are nonzero elements of M_0, M_2, M_3, M_4, M_5, we have $\dim M_k = 1$ for $k = 0, 2, 3, 4, 5$, which proves (ii).

Corollary 1.—*We have*

$$\dim M_k = \begin{cases} [k/6] & \text{if } k \equiv 1 \pmod 6,\ k \geqq 0 \\ [k/6] + 1 & \text{if } k \not\equiv 1 \pmod 6,\ k \geqq 0. \end{cases} \tag{21}$$

(Recall that $[x]$ denotes the *integral part* of x, i.e. the largest integer n such that $n \leqq x$.)

Formula (21) is true for $0 \leqq k < 6$ by (i) and (ii). Moreover, the two expressions increase by one unit when we replace k by $k+6$ (cf. (iii)). The formula is thus true for all $k \geqq 0$.

Corollary 2.—*The space M_k has for basis the family of monomials $G_2^\alpha G_3^\beta$ with α, β integers $\geqq 0$ and $2\alpha+3\beta = k$.*

We show first that these monomials *generate* M_k. This is clear for $k \leqq 3$ by (i) and (ii). For $k \geqq 4$ we argue by induction on k. Choose a pair (γ, δ) of integers $\geqq 0$ such that $2\gamma+3\delta = k$ (this is possible for all $k \geqq 2$). The modular form $g = G_2^\gamma G_3^\delta$ is not zero at infinity. If $f \in M_k$, there exists $\lambda \in \mathbf{C}$ such that $f-\lambda g$ is a cusp form, hence equal to Δh with $h \in M_{k-6}$, cf. (iii). One then applies the inductive hypothesis to h.

It remains to see that the above monomials are linearly independent; if they were not, the function G_2^3/G_3^2 would verify a nontrivial algebraic equation with coefficients in $\mathbf{C}$, thus would be constant, which is absurd because G_2 is zero at ρ but not G_3.

Remark.—Let $M = \sum_0^\infty M_k$ be the graded algebra which is the direct sum of the M_k and let $\varepsilon : \mathbf{C}[X, Y] \to M$ be the homomorphism which maps X on G_2 and Y on G_3. Cor. 2 is equivalent to saying that ε is an *isomorphism*. Hence, one can identify M with the polynomial algebra $\mathbf{C}[G_2, G_3]$.

3.3. *The modular invariant*

We put:

$$j = 1728 g_2^3/\Delta. \tag{22}$$

Proposition 5.—(a) *The function j is a modular function of weight* 0.

(b) *It is holomorphic in H and has a simple pole at infinity.*

(c) *It defines by passage to quotient a bijection of H/G onto* $\mathbf{C}$.

Assertion (a) comes from the fact that g_2^3 and Δ are both of weight 12; (b) comes from the fact that Δ is $\neq 0$ on H and has a simple zero at infinity, while g_2 is nonzero at infinity. To prove (c), one has to show that, if $\lambda \in \mathbf{C}$, the modular form $f_\lambda = 1728 g_2^3 - \lambda\Delta$ has a unique zero modulo G. To see this, one applies formula (20) with $f = f_\lambda$ and $k = 6$. The only decompositions of $k/6 = 1$ in the form $n+n'/2+n''/3$ with n, n', $n'' \geqq 0$ correspond to

$$(n, n', n'') = (1, 0, 0) \text{ or } (0, 2, 0) \text{ or } (0, 0, 3).$$

This shows that f_λ is zero at one and only one point of H/G.

Proposition 6.—*Let f be a meromorphic function on H. The following properties are equivalent:*

(i) *f is a modular function of weight* 0;

(ii) *f is a quotient of two modular forms of the same weight*;

(iii) *f is a rational function of j.*

The implications (iii) ⇒ (ii) ⇒ (i) are immediate. We show that (i) ⇒ (iii). Let f be a modular function. Being free to multiply f by a suitable polynomial in j, we can suppose that f is holomorphic on H. Since Δ is zero at infinity, there exists an integer $n \geqq 0$ such that $g = \Delta^n f$ is holomorphic at infinity. The function g is then a modular form of weight $12n$; by cor. 2 of theorem 4 we can write it as a linear combination of the $G_2^\alpha G_3^\beta$ with $2\alpha+3\beta = 6n$. By linearity, we are reduced to the case $g = G_2^\alpha G_3^\beta$, i.e. $f = G_2^\alpha G_3^\beta/\Delta^n$. But the relation $2\alpha+3\beta = 6n$ shows that $p = \alpha/2$ and $q = \beta/3$ are integers and one has $f = G_2^{3p} G_3^{2q}/\Delta^{p+q}$. Thus we are reduced to see that G_2^3/Δ and G_3^2/Δ are rational functions of j, which is obvious.

Remarks.—1) As stated above, it is possible to define in a natural way a structure of *complex analytic manifold* on the compactification $\widehat{H/G}$ of H/G. Prop. 5 means then that j defines an *isomorphism of* $\widehat{H/G}$ onto the Riemann sphere $\mathbf{S}_2 = \mathbf{C} \cup \{\infty\}$. As for prop. 6, it amounts to the well known fact that the only meromorphic functions on $\mathbf{S}_2$ are the rational functions.

2) The coefficient $1728 = 2^6 3^3$ has been introduced in order that j has a residue equal to 1 at infinity. More precisely, the series expansions of §4 show that:

$$j(z) = \frac{1}{q} + 744 + \sum_{n=1}^{\infty} c(n)q^n, \qquad z \in H, q = e^{2\pi iz}. \tag{23}$$

One has:

$$c(1) = 2^2 3^3\, 1823 = 196884, \quad c(2) = 2^{11} 5.2099 = 21493760.$$

The $c(n)$ are integers; they enjoy remarkable divisibility properties[1]:

$$\begin{aligned}
n &\equiv 0 \pmod{2^a} \Rightarrow c(n) \equiv 0 \pmod{2^{3a+8}} && \text{if } a \geq 1\\
n &\equiv 0 \pmod{3^a} \Rightarrow c(n) \equiv 0 \pmod{3^{2a+3}} && \text{''}\\
n &\equiv 0 \pmod{5^a} \Rightarrow c(n) \equiv 0 \pmod{5^{a+1}} && \text{''}\\
n &\equiv 0 \pmod{7^a} \Rightarrow c(n) \equiv 0 \pmod{7^a}\\
n &\equiv 0 \pmod{11^a} \Rightarrow c(n) \equiv 0 \pmod{11^a}.
\end{aligned}$$

§4. *Expansions at infinity*

4.1. *The Bernoulli numbers B_k*

They are defined by the power series expansion:[2]

---

[1] See on this subject A. O. L. ATKIN and J. N. O'BRIEN, Trans. Amer. Math. Soc., 126, 1967, as well as the paper of ATKIN in *Computers in mathematical research* (North Holland, 1968).

[2] In the literature, one also finds "Bernoulli numbers" b_k defined by

$$\frac{x}{e^x - 1} = \sum_{k=0}^{\infty} b_k x^k/k!\ ,$$

hence $b_0 = 1$, $b_1 = -1/2$, $b_{2k+1} = 0$ if $k > 1$, and $b_{2k} = (-1)^{k-1} B_k$. The b notation is better adapted to the study of congruence properties, and also to generalizations à la Leopoldt.

$$\frac{x}{e^x-1} = 1 - \frac{x}{2} + \sum_{k=1}^{\infty} (-1)^{k+1} B_k \frac{x^{2k}}{(2k)!}. \tag{24}$$

Numerical table

$$B_1 = \frac{1}{6},\ B_2 = \frac{1}{30},\ B_3 = \frac{1}{42},\ B_4 = \frac{1}{30},\ B_5 = \frac{5}{66},\ B_6 = \frac{691}{2730},$$

$$B_7 = \frac{7}{6},\ B_8 = \frac{3617}{510},\ B_9 = \frac{43867}{798},\ B_{10} = \frac{283.617}{330},\ B_{11} = \frac{11.131.593}{138},$$

$$B_{12} = \frac{103.2294797}{2730},\ B_{13} = \frac{13.657931}{6},\ B_{14} = \frac{7.9349.362903}{870}.$$

The B_k give the values of the Riemann zeta function for the positive even integers (and also for the negative odd integers):

Proposition 7.—*If k is an integer $\geqq 1$, then:*

$$\zeta(2k) = \frac{2^{2k-1}}{(2k)!} B_k \pi^{2k}. \tag{25}$$

The identity

$$z \operatorname{cotg} z = 1 - \sum_{k=1}^{\infty} B_k \frac{2^{2k} z^{2k}}{(2k)!} \tag{26}$$

follows from the definition of the B_k by putting $x = 2iz$. Moreover, taking the logarithmic derivative of

$$\sin z = z \prod_{n=1}^{\infty} \left(1 - \frac{z^2}{n^2\pi^2}\right), \tag{27}$$

we get:

$$z \operatorname{cotg} z = 1 + 2 \sum_{n=1}^{\infty} \frac{z^2}{z^2 - n^2\pi^2} \tag{28}$$

$$= 1 - 2 \sum_{n=1}^{\infty} \sum_{k=1}^{\infty} \frac{z^{2k}}{n^{2k}\pi^{2k}}.$$

Comparing (26) and (28), we get (25).

Examples $\zeta(2) = \dfrac{\pi^2}{2.3},\ \zeta(4) = \dfrac{\pi^4}{2.3^2.5},\ \zeta(6) = \dfrac{\pi^6}{3^3.5.7},$

$$\zeta(8) = \frac{\pi^8}{2.3^3.5^2.7},\ \zeta(10) = \frac{\pi^{10}}{3^5.5.7.11},\ \zeta(12) = \frac{691\pi^{12}}{3^6.5^3.7^2.11.13},$$

$$\zeta(14) = \frac{2\pi^{14}}{3^6.5^2.7.11.13}.$$

4.2. *Series expansions of the functions G_k*

We now give the Taylor expansion of the Eisenstein series $G_k(z)$ with respect to $q = e^{2\pi iz}$.

Let us start with the well known formula:

$$\pi \operatorname{cotg} \pi z = \frac{1}{z} + \sum_{m=1}^{\infty} \left(\frac{1}{z+m} + \frac{1}{z-m} \right). \tag{29}$$

We have on the other hand:

$$\pi \operatorname{cotg} \pi z = \pi \frac{\cos \pi z}{\sin \pi z} = i\pi \frac{q+1}{q-1} = i\pi - \frac{2i\pi}{1-q} = i\pi - 2i\pi \sum_{n=0}^{\infty} q^n, \tag{30}$$

Comparing, we get:

$$\frac{1}{z} + \sum_{m=1}^{\infty} \left(\frac{1}{z+m} + \frac{1}{z-m} \right) = i\pi - 2i\pi \sum_{n=0}^{\infty} q^n. \tag{31}$$

By successive differentiations of (31), we obtain the following formula (valid for $k \geqq 2$):

$$\sum_{m \in \mathbf{Z}} \frac{1}{(m+z)^k} = \frac{1}{(k-1)!} (-2i\pi)^k \sum_{n=1}^{\infty} n^{k-1} q^n. \tag{32}$$

Denote now by $\sigma_k(n)$ the sum $\sum_{d|n} d^k$ of kth-powers of positive divisors of n.

Proposition 8.—*For every integer $k \geqq 2$, one has:*

$$G_k(z) = 2\zeta(2k) + 2 \frac{(2i\pi)^{2k}}{(2k-1)!} \sum_{n=1}^{\infty} \sigma_{2k-1}(n) q^n. \tag{33}$$

We expand:

$$G_k(z) = \sum_{(n,m) \neq (0,0)} \frac{1}{(nz+m)^{2k}}$$

$$= 2\zeta(2k) + 2 \sum_{n=1}^{\infty} \sum_{m \in \mathbf{Z}} \frac{1}{(nz+m)^{2k}}.$$

Applying (32) with z replaced by nz, we get

$$G_k(z) = 2\zeta(2k) + \frac{2(-2\pi i)^{2k}}{(2k-1)!} \sum_{d=1}^{\infty} \sum_{a=1}^{\infty} d^{2k-1} q^{ad}$$

$$= 2\zeta(2k) + \frac{2(2\pi i)^{2k}}{(2k-1)!} \sum_{n=1}^{\infty} \sigma_{2k-1}(n) q^n.$$

Corollary.—*$G_k(z) = 2\zeta(2k) E_k(z)$ with*

$$E_k(z) = 1 + \gamma_k \sum_{n=1}^{\infty} \sigma_{2k-1}(n) q^n \tag{34}$$

and

$$\gamma_k = (-1)^k \frac{4k}{B_k}. \tag{35}$$

One defines $E_k(z)$ as the quotient of $G_k(z)$ by $2\zeta(2k)$; it is clear that $E_k(z)$ is given by (34). The coefficient γ_k is computed using prop. 7:

$$\gamma_k = \frac{(2i\pi)^{2k}}{(2k-1)!}\frac{1}{\zeta(2k)} = \frac{(2\pi)^{2k}(-1)^k}{(2k-1)!}\frac{(2k)!}{2^{2k-1}B_k\pi^{2k}} = (-1)^k\frac{4k}{B_k}.$$

Examples

$$E_2 = 1+240\sum_{n=1}^{\infty}\sigma_3(n)q^n, \qquad g_2 = (2\pi)^4\frac{1}{2^2.3}E_2$$

$$E_3 = 1-504\sum_{n=1}^{\infty}\sigma_5(n)q^n, \qquad g_3 = (2\pi)^6\frac{1}{2^3.3^3}E_3$$

$$E_4 = 1+480\sum_{n=1}^{\infty}\sigma_7(n)q^n \qquad (480 = 2^5.3.5)$$

$$E_5 = 1-264\sum_{n=1}^{\infty}\sigma_9(n)q^n \qquad (264 = 2^3.3.11)$$

$$E_6 = 1+\frac{65520}{691}\sum_{n=1}^{\infty}\sigma_{11}(n)q^n \qquad (65520 = 2^4.3^2.5.7.13)$$

$$E_7 = 1-24\sum_{n=1}^{\infty}\sigma_{13}(n)q^n.$$

Remark.—We have seen in n° 3.2 that the space of modular forms of weight 8 (resp. 10) is of dimension 1. Hence:

$$E_2^2 = E_4, \quad E_2E_3 = E_5. \tag{36}$$

This is equivalent to the identities:

$$\sigma_7(n) = \sigma_3(n)+120\sum_{m=1}^{n-1}\sigma_3(m)\sigma_3(n-m)$$

$$11\sigma_9(n) = 21\sigma_5(n)-10\sigma_3(n)+5040\sum_{m=1}^{n-1}\sigma_3(n)\sigma_5(n-m).$$

More generally, every E_k can be expressed as a *polynomial* in E_2 and E_3.

4.3. *Estimates for the coefficients of modular forms*

Let

$$f(z) = \sum_{n=0}^{\infty}a_nq^n \quad (q = e^{2\pi iz}) \tag{37}$$

be a modular form of weight $2k$, $k \geqq 2$. We are interested in the growth of the a_n:

Proposition 9.—*If $f = G_k$, the order of magnitude of a_n is n^{2k-1}. More precisely, there exist two constants A, $B > 0$ such that*

$$An^{2k-1} \leqq |a_n| \leqq Bn^{2k-1}. \tag{38}$$

Prop. 8 shows that there exists a constant $A > 0$ such that

$$a_n = (-1)^k A\sigma_{2k-1}(n), \qquad \text{hence } |a_n| = A\sigma_{2k-1}(n) \geqq An^{2k-1}.$$

On the other hand:

$$\frac{|a_n|}{n^{2k-1}} = A \sum_{d|n} \frac{1}{d^{2k-1}} \leqq A \sum_{d=1}^{\infty} \frac{1}{d^{2k-1}} = A\zeta(2k-1) < +\infty.$$

Theorem 5 (Hecke).—*If f is a cusp form of weight $2k$, then*

$$a_n = O(n^k). \tag{39}$$

(In other words, the quotient $\dfrac{|a_n|}{n^k}$ remains bounded when $n \to \infty$.)

Because f is a cusp form, we have $a_0 = 0$ and can factor q out of the expansion (37) of f. Hence:

$$|f(z)| = O(q) = O(e^{-2\pi y}) \quad \text{with } y = Im(z), \quad \text{when } q \text{ tends to } 0. \tag{40}$$

Let $\phi(z) = |f(z)|y^k$. Formulas (1) and (2) show that ϕ is *invariant* under the modular group G. In addition, ϕ is continuous on the fundamental domain D and formula (40) shows that ϕ tends to 0 for $y \to \infty$. This implies that ϕ is *bounded*, i.e. there exists a constant M such that

$$|f(z)| \leqq My^{-k} \qquad \text{for } z \in H. \tag{41}$$

Fix y and vary x between 0 and 1. The point $q = e^{2\pi i(x+iy)}$ runs along a circle C_y of center 0. By the residue formula,

$$a_n = \frac{1}{2\pi i}\int_{C_y} f(z)q^{-n-1}dq = \int_0^1 f(x+iy)q^{-n}dx.$$

(One could also deduce this formula from that giving the Fourier coefficients of a periodic function.)

Using (41), we get from this

$$|a_n| \leqq My^{-k}e^{2\pi ny}.$$

This inequality is valid for all $y > 0$. For $y = 1/n$, it gives $|a_n| \leqq e^{2\pi}Mn^k$. The theorem follows from this.

Corollary.—*If f is not a cusp form, then the order of magnitude of a_n is n^{2k-1}.*

We write f in the form $\lambda G_k + h$ with $\lambda \neq 0$ and a cusp form h and we apply prop. 9 and th. 5, taking into account the fact that n^k is "negligible" compared to n^{2k-1}.

Remark.—The exponent k of theorem 5 can be improved. Indeed, Deligne has shown (cf. 5.6.3 below) that

$$a_n = O(n^{k-1/2}\sigma_0(n)),$$

where $\sigma_0(n)$ is the number of positive divisors of n. This implies that

$$a_n = O(n^{k-1/2+\varepsilon}) \qquad \text{for every } \varepsilon > 0.$$

4.4. *Expansion of* Δ

Recall that

$$\begin{aligned}\Delta &= g_2^3 - 27g_3^2 = (2\pi)^{12}2^{-6}3^{-3}(E_2^3 - E_3^2) \\ &= (2\pi)^{12}(q - 24q^2 + 252q^3 - 1472q^4 + \ldots).\end{aligned} \tag{42}$$

Theorem 6 (Jacobi).—$\Delta = (2\pi)^{12} q \prod_{n=1}^{\infty} (1-q^n)^{24}$.

[This formula is proved in the most natural way by using elliptic functions. Since this method would take us too far afield, we sketch below a different proof, which is "elementary" but somewhat artificial; for more details, the reader can look into A. HURWITZ, *Math. Werke*, Bd. I, pp. 578–595.]

We put:

$$F(z) = q \prod_{n=1}^{\infty} (1-q^n)^{24}. \tag{43}$$

To prove that F and Δ are proportional, it suffices to show that F is a modular form of weight 12; indeed, the fact that the expansion of G has constant term zero will show that F is a cusp form and we know (th. 4) that the space M_6^0 of cusp forms of weight 12 is of dimension 1. By prop. 1 of n° 2.1, all there is to do is to prove that:

$$F(-1/z) = z^{12}F(z). \tag{44}$$

We use for this the double series

$$G_1(z) = \sum_n \sum_m{}' \frac{1}{(m+nz)^2}, \quad G(z) = \sum_m \sum_n{}' \frac{1}{(m+nz)^2}$$

$$H_1(z) = \sum_n \sum_m{}' \frac{1}{(m-1+nz)(m+nz)}, \quad H(z) = \sum_m \sum_n{}' \frac{1}{(m-1+nz)(m+nz)}$$

where the sign Σ' indicates that (m,n) runs through all $m \in \mathbf{Z}$, $n \in \mathbf{Z}$ with $(m,n) \neq (0,0)$ for G and G_1 and $(m,n) \neq (0,0), (1,0)$ for H and H_1. (Notice the order of the summations!)

The series H_1 and H are easy to calculate explicitly because of the formula:

$$\frac{1}{(m-1+nz)(m+nz)} = \frac{1}{m-1+nz} - \frac{1}{m+nz}.$$

One finds that they converge, and that

$$H_1 = 2, \; H = 2-2\pi i/z.$$

Moreover, the double series with general term

$$\frac{1}{(m-1+nz)(m+nz)} - \frac{1}{(m+nz)^2} = \frac{1}{(m+nz)^2(m-1+nz)}$$

is absolutely summable. This shows that $G_1 - H_1$ and $G - H$ coincide, thus that the series G and G_1 converge (with order of summation indicated) and that

$$G_1(z) - G(z) = H_1(z) - H(z) = \frac{2\pi i}{z}.$$

It is clear moreover that $G_1(-1/z) = z^2 G(z)$. Hence:

$$G_1(-1/z) = z^2 G_1(z) - 2\pi i z. \tag{45}$$

On the other hand, a computation similar to that of prop. 8 gives

$$G_1(z) = \frac{\pi^2}{3} - 8\pi^2 \sum_{n=1}^{\infty} \sigma_1(n) q^n. \tag{46}$$

Now, go back to the function F defined by (43). Its logarithmic differential is

$$\frac{dF}{F} = \frac{dq}{q}\left(1 - 24 \sum_{n,m=1}^{\infty} n q^{nm}\right) = \frac{dq}{q}\left(1 - 24 \sum_{n=1}^{\infty} \sigma_1(n) q^n\right). \tag{47}$$

By comparing with (46), we get:

$$\frac{dF}{F} = \frac{6i}{\pi} G_1(z) dz. \tag{48}$$

Combining (45) and (48), we have

$$\begin{aligned} \frac{dF(-1/z)}{F(-1/z)} &= \frac{6i}{\pi} G_1(-1/z) \frac{dz}{z^2} = \frac{6i}{\pi} \frac{dz}{z^2} (z^2 G_1(z) - 2\pi i z) \\ &= \frac{dF(z)}{F(z)} + 12 \frac{dz}{z}. \end{aligned} \tag{49}$$

Thus the two functions $F(-1/z)$ and $z^{12}F(z)$ have the same logarithmic differential. Hence there exists a constant k such that $F(-1/z) = kz^{12}F(z)$ for all $z \in H$. For $z = i$, we have $z^{12} = 1$, $-1/z = z$ and $F(z) \neq 0$; this shows that $k = 1$, which proves (44), q.e.d.

Remark.—One finds another "elementary" proof of identity (44) in C. L. Siegel, *Gesamm. Abh.*, III, n° 62. See also *Seminar on complex multiplication*, III, §6.

4.5. *The Ramanujan function*

We denote by $\tau(n)$ the nth coefficient of the cusp form $F(z) = (2\pi)^{-12}\Delta(z)$. Thus

$$(50) \qquad \sum_{n=1}^{\infty} \tau(n)q^n = q\prod_{n=1}^{\infty}(1-q^n)^{24}.$$

The function $n \mapsto \tau(n)$ is called the *Ramanujan function*.

Numerical table [1]

$\tau(1) = 1$, $\tau(2) = -24$, $\tau(3) = 252$, $\tau(4) = -1472$, $\tau(5) = 4830$,
$\tau(6) = -6048$, $\tau(7) = -16744$, $\tau(8) = 84480$, $\tau(9) = -113643$,
$\tau(10) = -115920$, $\tau(11) = 534612$, $\tau(12) = -370944$.

Properties of $\tau(n)$

$$(51) \qquad \tau(n) = O(n^6),$$

because Δ is of weight 12, cf. n° 4.3, th. 5. (By Deligne's theorem, we even have $\tau(n) = O(n^{11/2+\varepsilon})$ for every $\varepsilon > 0$.)

$$(52) \qquad \tau(nm) = \tau(n)\tau(m) \quad \text{if } (n, m) = 1$$

$$(53) \quad \tau(p^{n+1}) = \tau(p)\tau(p^n) - p^{11}\tau(p^{n-1}) \quad \text{for } p \text{ prime}, n > 1, \text{ cf. n° 5.5. below.}$$

The identities (52) and (53) were conjectured by Ramanujan and first proved by Mordell. One can restate them by saying that the Dirichlet series $L_\tau(s) = \sum_{n=1}^{\infty} \tau(n)/n^s$ has the following eulerian expansion:

$$(54) \qquad L_\tau(s) = \prod_{p \in P} \frac{1}{1-\tau(p)p^{-s}+p^{11-2s}}, \quad \text{cf. n° 5.4.}$$

By a theorem of Hecke (cf. n° 5.4) the function L_τ extends to an entire function in the complex plane and the function

$$(2\pi)^{-s}\Gamma(s)L_\tau(s)$$

is invariant by $s \mapsto 12-s$.

The $\tau(n)$ enjoy various congruences modulo 2^{12}, 3^6, 5^3, 7, 23, 691. We quote some special cases (without proof):

$$(55) \qquad \tau(n) \equiv n^2\sigma_7(n) \pmod{3^3}$$

$$(56) \qquad \tau(n) \equiv n\sigma_3(n) \pmod 7$$

$$(57) \qquad \tau(n) \equiv \sigma_{11}(n) \pmod{691}.$$

For other examples, and their interpretation in terms of "l-adic representations" see *Sém. Delange-Pisot-Poitou* 1967/68, exposé 14, *Sém. Bourbaki* 1968/69, exposé 355 and Swinnerton-Dyer's lecture at Antwerp (*Lecture Notes*, n° 350, Springer, 1973).

---

[1] This table is taken from D. H. LEHMER, *Ramanujan's function* $\tau(n)$, *Duke Math. J.*, 10, 1943, which gives the values of $\tau(n)$ for $n \leq 300$.

We end up with an *open question*, raised by D. H. Lehmer:
Is it true that $\tau(n) \neq 0$ for all $n \geq 1$?
It is so for $n \leq 10^{15}$.

§5. *Hecke operators*

5.1. *Definition of the* $T(n)$

Correspondences.—Let E be a set and let X_E be the free abelian group generated by E. A *correspondence* on E (with integer coefficients) is a homomorphism T of X_E into itself. We can give T by its values on the elements x of E:

$$T(x) = \sum_{y \in E} n_y(x) y, \quad n_y(x) \in \mathbf{Z}, \tag{58}$$

the $n_y(x)$ being zero for almost all y.

Let F be a numerical valued function on E. By $\mathbf{Z}$-linearity it extends to a function, again denoted F, on X_E. The transform of F by T, denoted TF, is the restriction to E of the function $F \circ T$. With the notations of (58),

$$TF(x) = F(T(x)) = \sum_{y \in E} n_y(x) F(y). \tag{59}$$

The $T(n)$.—Let $\mathscr{R}$ be the set of lattices of $\mathbf{C}$ (see n° 2.2). Let n be an integer $\geqq 1$. We denote by $T(n)$ the correspondence on $\mathscr{R}$ which transforms a lattice to the sum (in $X_{\mathscr{R}}$) of its sub-lattices of index n. Thus we have:

$$T(n)\Gamma = \sum_{(\Gamma : \Gamma') = n} \Gamma' \quad \text{if } \Gamma \in \mathscr{R}. \tag{60}$$

The sum on the right side is finite. Indeed, the lattices Γ' all contain $n\Gamma$ and their number is also the number of subgroups of order n of $\Gamma/n\Gamma = (\mathbf{Z}/n\mathbf{Z})^2$. If n is prime, one sees easily that this number is equal to $n+1$ (number of points of the projective line over a field with n elements).

We also use the homothety operators R_λ ($\lambda \in \mathbf{C}^*$) defined by

$$R_\lambda \Gamma = \lambda \Gamma \quad \text{if } \Gamma \in \mathscr{R}. \tag{61}$$

Formulas.—It makes sense to compose the correspondences $T(n)$ and R_λ, since they are endomorphisms of the abelian group $X_{\mathscr{R}}$.

Proposition 10.—*The correspondences* $T(n)$ *and* R_λ *verify the identities*

$$R_\lambda R_\mu = R_{\lambda\mu} \qquad (\lambda, \mu \in \mathbf{C}^*) \tag{62}$$

$$R_\lambda T(n) = T(n) R_\lambda \qquad (n \geqq 1, \lambda \in \mathbf{C}^*) \tag{63}$$

$$T(m)T(n) = T(mn) \qquad \textit{if } (m, n) = 1 \tag{64}$$

$$T(p^n)T(p) = T(p^{n+1}) + pT(p^{n-1})R_p \qquad (p \textit{ prime}, n \geqq 1). \tag{65}$$

Formulas (62) and (63) are trivial.

Formula (64) is equivalent to the following assertion: Let m, n be two relatively prime integers $\geqq 1$, and let Γ'' be a sublattice of a lattice Γ of index mn; there exists a unique sublattice Γ' of Γ, containing Γ'', such that $(\Gamma:\Gamma') = n$ and $(\Gamma':\Gamma'') = m$. This assertion follows itself from the fact that the group Γ/Γ'', which is of order mn, decomposes uniquely into a direct sum of a group of order m and a group of order n (Bezout's theorem).

To prove (65), let Γ be a lattice. Then $T(p^n)T(p)\Gamma$, $T(p^{n+1})\Gamma$ and $T(p^{n-1})R_p\Gamma$ are linear combinations of lattices contained in Γ and of index p^{n+1} in Γ (note that $R_p\Gamma$ is of index p^2 in Γ). Let Γ'' be such a lattice; in the above linear combinations it appears with coefficients a, b, c, say; we have to show that $a = b+pc$, i.e. that $a = 1+pc$ since b is clearly equal to 1.

We have two cases:

i) Γ'' is not contained in $p\Gamma$. Then $c = 0$ and a is the number of lattices Γ', intermediate between Γ and Γ'', and of index p in Γ; such a lattice Γ' contains $p\Gamma$. In $\Gamma/p\Gamma$ the image of Γ' is of index p and it contains the image of Γ'' which is of order p (hence also of index p because $\Gamma/p\Gamma$ is of order p^2); hence there is only one Γ' which does the trick. This gives $a = 1$ and the formula $a = 1+pc$ is valid.

ii) $L'' \subset p\Gamma$. We have $c = 1$; any lattice Γ' of index p in Γ contains $p\Gamma$, thus *a fortiori* Γ''. This gives $a = p+1$ and $a = 1+pc$ is again valid.

Corollary 1.—*The* $T(p^n)$, $n > 1$, *are polynomials in* $T(p)$ *and* R_p.
This follows from (65) by induction on n.

Corollary 2.—*The algebra generated by the* R_λ *and the* $T(p)$, p *prime, is commutative; it contains all the* $T(n)$.
This follows from prop. 10 and cor. 1.

Action of $T(n)$ *on the functions of weight* $2k$.
Let F be a function on $\mathcal{R}$ of weight $2k$ (cf. n° 2.2). By definition

$$R_\lambda F = \lambda^{-2k}F \quad \text{for all } \lambda \in \mathbf{C}^*. \tag{66}$$

Let n be an integer $\geqq 1$. Formula (63) shows that

$$R_\lambda(T(n)F) = T(n)(R_\lambda F) = \lambda^{-2k}T(n)F,$$

in other words $T(n)F$ is also of weight $2k$. Formulas (64) and (65) give:

$$T(m)T(n)F = T(mn)F \qquad \text{if } (m, n) = 1, \tag{67}$$

$$T(p)T(p^n)F = T(p^{n+1})F+p^{1-2k}T(p^{n-1})F, \qquad p \text{ prime}, n \geqq 1. \tag{68}$$

5.2. *A matrix lemma*

Let Γ be a lattice with basis $\{\omega_1, \omega_2\}$ and let n be an integer $\geqq 1$. The following lemma gives all the sublattices of Γ of index n:

Lemma 2.—*Let* S_n *be the set of integer matrixes* $\begin{pmatrix} a & b \\ 0 & d \end{pmatrix}$ *with* $ad = n$, $a \geqq 1$, $0 \leqq b < d$. *If* $\sigma = \begin{pmatrix} a & b \\ 0 & d \end{pmatrix}$ *is contained in* S_n, *let* Γ_σ *be the sublattice* *of Γ having for basis*

$$\omega_1' = a\omega_1 + b\omega_2,\ \omega_2' = d\omega_2.$$

The map $\sigma \mapsto \Gamma_\sigma$ is a bijection of S_n onto the set $\Gamma(n)$ of sublattices of index n in Γ.

The fact that Γ_σ belongs to $\Gamma(n)$ follows from the fact that $\det(\sigma) = n$. Conversely let $\Gamma' \in \Gamma(n)$. We put

$$Y_1 = \Gamma/(\Gamma' + \mathbf{Z}\omega_2) \quad \text{and} \quad Y_2 = \mathbf{Z}\omega_2/(\Gamma' \cap \mathbf{Z}\omega_2).$$

These are cyclic groups generated respectively by the images of ω_1 and ω_2. Let a and d be their orders. The exact sequence

$$0 \to Y_2 \to \Gamma/\Gamma' \to Y_1 \to 0$$

shows that $ad = n$. If $\omega_2' = d\omega_2$, then $\omega_2' \in \Gamma'$. On the other hand, there exists $\omega_1' \in \Gamma'$ such that

$$\omega_1' \equiv a\omega_1 \pmod{\mathbf{Z}\omega_2}.$$

It is clear that ω_1' and ω_2' form a basis of Γ'. Moreover, we can write ω_1' in the form

$$\omega_1' = a\omega_1 + b\omega_2 \quad \text{with } b \in \mathbf{Z},$$

where b is uniquely determined modulo d. If we impose on b the inequality $0 \leqq b < d$, this fixes b, thus also ω_1'. Thus we have associated to every $\Gamma' \in \Gamma(n)$ a matrix $\sigma(\Gamma') \in S_n$, and one checks that the maps $\sigma \mapsto \Gamma_\sigma$ and $\Gamma' \mapsto \sigma(\Gamma')$ are inverses to each other; the lemma follows.

Example.—If p is a prime, the elements of S_p are the matrix $\begin{pmatrix} p & 0 \\ 0 & 1 \end{pmatrix}$ and the p matrices $\begin{pmatrix} 1 & b \\ 0 & p \end{pmatrix}$ with $0 \leqq b < p$.

5.3. *Action of $T(n)$ on modular functions*

Let k be an integer, and let f be a weakly modular function of weight $2k$, cf. n° 2.1. As we saw in n° 2.2, f corresponds to a function F of weight $2k$ on $\mathcal{R}$ such that

$$F(\Gamma(\omega_1, \omega_2)) = \omega_2^{-2k} f(\omega_1/\omega_2). \tag{69}$$

We define $T(n)f$ as *the function on H associated to the function $n^{2k-1}T(n)F$ on $\mathcal{R}$.* (Note the numerical coefficient n^{2k-1} which gives formulas "without denominators" in what follows.) Thus by definition:

$$T(n)f(z) = n^{2k-1}T(n)F(\Gamma(z, 1)), \tag{70}$$

or else by lemma 2:

$$T(n)f(z) = n^{2k-1} \sum_{\substack{a \geqq 1,\, ad = n \\ 0 \leqq b < d}} d^{-2k} f\left(\frac{az+b}{d}\right). \tag{71}$$

Proposition 11.—*The function $T(n)f$ is weakly modular of weight $2k$. It is holomorphic on H if f is. We have*:

$$T(m)T(n)f = T(mn) \quad \text{if} \quad (m,n) = 1, \tag{72}$$

$$T(p)T(p^n)f = T(p^{n+1})f + p^{2k-1}T(p^{n-1})f, \quad \text{if } p \text{ is prime}, n \geqq 1. \tag{73}$$

Formula (71) shows that $T(n)f$ is meromorphic on H, thus weakly modular; if in addition f is holomorphic, so is $T(n)f$. Formulas (72) and (73) follow from formulas (67) and (68) taking into account the numerical coefficient n^{2k-1} incorporated into the definition of $T(n)f$.

Behavior at infinity.—We suppose that f is a *modular function*, i.e. is meromorphic at infinity. Let

$$f(z) = \sum_{m \in \mathbf{Z}} c(m)q^m \tag{74}$$

be its Laurent expansion with respect to $q = e^{2\pi iz}$.

Proposition 12.—*The function $T(n)f$ is a modular function. We have*

$$T(n)f(z) = \sum_{m \in \mathbf{Z}} \gamma(m)q^m \tag{75}$$

with

$$\gamma(m) = \sum_{\substack{a|(n,m)\\ a \geqq 1}} a^{2k-1}c\left(\frac{mn}{a^2}\right). \tag{76}$$

By definition, we have:

$$T(n)f(z) = n^{2k-1} \sum_{\substack{ad=n,\ a \geqq 1\\ 0 \leqq b < d}} d^{-2k} \sum_{m \in \mathbf{Z}} c(m)e^{2\pi im(az+b)/d}$$

Now the sum

$$\sum_{0 \leqq b < d} e^{2\pi i\, bm/d}$$

is equal to d if d divides m and to 0 otherwise. Thus we have, putting $m/d = m'$:

$$T(n)f(z) = n^{2k-1} \sum_{\substack{ad=n\\ a \geqq 1,\ m' \in \mathbf{Z}}} d^{-2k+1}c(m'd)q^{am'}.$$

Collecting powers of q, this gives:

$$T(n)f(z) = \sum_{\mu \in \mathbf{Z}} q^\mu \sum_{\substack{a|(n,\mu)\\ a \geqq 1}} \left(\frac{n}{d}\right)^{2k-1} c\left(\frac{\mu d}{a}\right).$$

Since f is meromorphic at infinity, there exists an integer $N \geqq 0$ such that $c(m) = 0$ if $m \leqq -N$. The $c\left(\frac{\mu d}{a}\right)$ are thus zero for $\mu \leqq -nN$, which shows that $T(n)f$ is also meromorphic at infinity. Since it is weakly modular, it is a modular function. The fact that its coefficients are given by formula (76) follows from the above computation.

Corollary 1.—$\gamma(0) = \sigma_{2k-1}(n)c(0)$ *and* $\gamma(1) = c(n)$.

Corollary 2.—*If* $n = p$ *with* p *prime, one has*

$$\gamma(m) = c(pm) \quad \text{if } m \not\equiv 0 \pmod p$$

$$\gamma(m) = c(pm)+p^{2k-1}c(m/p) \quad \text{if } m \equiv 0 \pmod p.$$

Corollary 3.—*If* f *is a modular form (resp. a cusp form), so is* $T(n)f$.
This is clear.

Thus, the $T(n)$ *act* on the spaces M_k and M_k^0 of n° 3.2. As we saw above, the operators thus defined *commute* with each other and satisfy the identities:

$$T(m)T(n) = T(mn) \quad \text{if } (m, n) = 1 \tag{72}$$

$$T(p)T(p^n) = T(p^{n+1})+p^{2k-1}T(p^{n-1}) \quad \text{if } p \text{ is prime, } n \geqq 1. \tag{73}$$

5.4. *Eigenfunctions of the* $T(n)$

Let $f(z) = \sum_{n=0}^{\infty} c(n)q^n$ be a modular form of weight $2k$, $k > 0$, not identically zero. We assume that f is an *eigenfunction* of all the $T(n)$, i.e. that there exists a complex number $\lambda(n)$ such that

$$T(n)f = \lambda(n)f \quad \text{for all } n \geqq 1. \tag{77}$$

Theorem 7.—a) *The coefficient* $c(1)$ *of* q *in* f *is* $\neq 0$.
b) *If* f *is normalized by the condition* $c(1) = 1$, *then*

$$c(n) = \lambda(n) \quad \text{for all } n > 1. \tag{78}$$

Cor. 1 to prop. 12 shows that the coefficient of q in $T(n)f$ is $c(n)$. On the other hand, by (77), it is also $\lambda(n)c(1)$. Thus we have $c(n) = \lambda(n)c(1)$. If $c(1)$ were zero, all the $c(n)$, $n > 0$, would be zero, and f would be a constant which is absurd. Hence a) and b).

Corollary 1.—*Two modular forms of weight* $2k$, $k > 0$, *which are eigenfunctions of the* $T(n)$ *with the same eigenvalues* $\lambda(n)$, *and which are normalized, coincide.*
This follows from a) applied to the difference of the two functions.

Corollary 2.—*Under the hypothesis of theorem* 7, b):

$$c(m)c(n) = c(mn) \quad \text{if } (m, n) = 1 \tag{79}$$

$$c(p)c(p^n) = c(p^{n+1})+p^{2k-1}c(p^{n-1}). \tag{80}$$

Indeed the eigenvalues $\lambda(n) = c(n)$ satisfy the same identities (72) and (73) as the $T(n)$.
Formulas (79) and (80) can be translated analytically in the following manner:

Let

$$\Phi_f(s) = \sum_{n=1}^{\infty} c(n)/n^s \tag{81}$$

be the Dirichlet series defined by the $c(n)$; by the cor. of th. 5, this series converges absolutely for $R(s) > 2k$.

Corollary 3.—*We have*:

$$\Phi_f(s) = \prod_{p\in P} \frac{1}{1-c(p)p^{-s}+p^{2k-1-2s}} \tag{82}$$

By (79) the function $n \mapsto c(n)$ is multiplicative. Thus lemma 4 of chap. VII, n° 3.1 shows that $\Phi_f(s)$ is the product of the series $\sum_{n=0}^{\infty} c(p^n)p^{-ns}$. Putting $p^{-s} = T$, we are reduced to proving the identity

$$\sum_{n=0}^{\infty} c(p^n)T^n = \frac{1}{\Phi_{f,p}(T)} \quad \text{where} \quad \Phi_{f,p}(T) = 1-c(p)T+p^{2k-1}T^2. \tag{83}$$

Form the series

$$\psi(T) = \left(\sum_{n=0}^{\infty} c(p^n)T^n\right)(1-c(p)T+p^{2k-1}T^2).$$

The coefficient of T in ψ is $c(p)-c(p) = 0$. That of T^{n+1}, $n \geqq 1$, is

$$c(p^{n+1})-c(p)c(p^n)+p^{2k-1}c(p^{n-1}),$$

which is zero by (80). Thus the series ψ is reduced to its constant term $c(1) = 1$, and this proves (83).

Remarks.—1) Conversely, formulas (81) and (82) imply (79) and (80).

2) Hecke has proved that Φ_f extends analytically to a meromorphic function on the whole complex plane (it is even holomorphic if f is a cusp form) and that the function

$$X_f(s) = (2\pi)^{-s}\Gamma(s)\Phi_f(s) \tag{84}$$

satisfies *the functional equation*

$$X_f(s) = (-1)^k X_f(2k-s). \tag{85}$$

The proof uses *Mellin's formula*

$$X_f(s) = \int_0^{\infty} (f(iy)-f(\infty))y^s \frac{dy}{y}$$

combined with the identity $f(-1/z) = z^{2k}f(z)$. Hecke also proved a converse: every Dirichlet series Φ which satisfies a functional equation of this type, and some regularity and growth hypothesis, comes from a modular form f of weight $2k$; moreover, f is a normalized eigenfunction of the $T(n)$

if and only if ϕ is an Eulerian product of type (82). See for more details E. HECKE, *Math. Werke*, n° 33, and A. WEIL, *Math. Annalen*, 168, 1967.

5.5. *Examples*

a) *Eisenstein series.*—Let k be an integer $\geqq 2$.

Proposition 13.—*The Eisenstein series G_k is an eigenfunction of $T(n)$; the corresponding eigenvalue is $\sigma_{2k-1}(n)$ and the normalized eigenfunction is*

$$(86)\qquad (-1)^k \frac{B_k}{4k} E_k = (-1)^k \frac{B_k}{4k} + \sum_{n=1}^{\infty} \sigma_{2k-1}(n)q^n.$$

The corresponding Dirichlet series is $\zeta(s)\zeta(s-2k+1)$.

We prove first that G_k is an eigenfunction of $T(n)$; it suffices to do this for $T(p)$, p prime. Consider G_k as a function on the set $\mathscr{R}$ of lattices of $\mathbf{C}$; we have:

$$G_k(\Gamma) = \sum_{\gamma\in\Gamma}' 1/\gamma^{2k}, \quad \text{cf. n° 2.3},$$

and

$$T(p)G_k(\Gamma) = \sum_{(\Gamma:\Gamma')=p} \sum_{\gamma\in\Gamma'}' 1/\gamma^{2k}.$$

Let $\gamma \in \Gamma$. If $\gamma \in p\Gamma$ then γ belongs to each of the $p+1$ sublattices of Γ of index p; its contribution in $T(p)G_k(\Gamma)$ is $(p+1)/\gamma^{2k}$. If $\gamma \in \Gamma - p\Gamma$, then γ belongs to only one sublattice of index p and its contribution is $1/\gamma^{2k}$. Thus

$$\begin{aligned} T(p)G_k(\Gamma) &= G_k(\Gamma)+p\sum_{\gamma\in p\Gamma} 1/\gamma^{2k} = G_k(\Gamma)+pG_k(p\Gamma) \\ &= (1+p^{1-2k})G_k(\Gamma), \end{aligned}$$

which proves that G_k (viewed as a *function on* $\mathscr{R}$) is an eigenfunction of $T(p)$ with eigenvalue $1+p^{1-2k}$; viewed as a modular form, G_k is thus an eigenfunction of $\Gamma(p)$ with eigenvalue $p^{2k-1}(1+p^{1-2k}) = \sigma_{2k-1}(p)$. Formulas (34) and (35) of n° 4.2 show that the *normalized* eigenfunction associated with G_k is

$$(-1)^k \frac{B_k}{4k} + \sum_{n=1}^{\infty} \sigma_{2k-1}(n)q^n.$$

This also shows that the eigenvalues of $T(n)$ are $\sigma_{2k-1}(n)$. Finally

$$\begin{aligned} \sum_{n=1}^{\infty} \sigma_{2k-1}(n)/n^s &= \sum_{a,d\geqq 1} a^{2k-1}/a^s d^s \\ &= \Big(\sum_{d\geqq 1} 1/d^s\Big)\Big(\sum_{a\geqq 1} 1/a^{s+1-2k}\Big) \\ &= \zeta(s)\zeta(s-2k+1). \end{aligned}$$

b) *The Δ function*

Proposition 14.—*The Δ function is an eigenfunction of $T(n)$. The corresponding eigenvalue is $\tau(n)$ and the normalized eigenfunction is*

$$(2\pi)^{-12}\Delta = q\prod_{n=1}^{\infty}(1-q^n)^{24} = \sum_{n=1}^{\infty} \tau(n)q^n.$$

This is clear, since the space of cusp forms of weight 12 is of dimension 1, and is stable by the $T(n)$.

Corollary.—*We have*

(52) $$\tau(nm) = \tau(n)\tau(m) \qquad \textit{if } (n, m) = 1,$$

(53) $$\tau(p)\tau(p^n) = \tau(p^{n+1})+p^{11}\tau(p^{n-1}) \qquad \textit{if } p \textit{ is a prime, } n \geqq 1.$$

This follows from cor. 2 of th. 7.

Remark.—There are similar results when the space M_k^0 of cusp forms of weight $2k$ has dimension 1; this happens for

$$k = 6, 8, 9, 10, 11, 13 \text{ with basis } \Delta, \Delta G_2, \Delta G_3, \Delta G_4, \Delta G_5, \text{ and } \Delta G_7.$$

5.6. *Complements*

5.6.1. *The Petersson scalar product.*

Let f, g be two cusp forms of weight $2k$ with $k > 0$. One proves easily that the measure

$$\mu(f,g) = f(z)\overline{g(z)}y^{2k}dxdy/y^2 \qquad (x = R(z), y = Im(z))$$

is *invariant* by G and that it is a *bounded* measure on the quotient space H/G. By putting

(87) $$\langle f, g\rangle = \int_{H/G} \mu(f, g) = \int_D f(z)\overline{g(z)}y^{2k-2}dxdy,$$

we obtain a hermitian scalar product on M_k^0 which is *positive* and *non-degenerate.* One can check that

(88) $$\langle T(n)f, g\rangle = \langle f, T(n)g\rangle,$$

which means that the $T(n)$ are *hermitian* operators with respect to $\langle f, g\rangle$. Since the $T(n)$ commute with each other, a well known argument shows that *there exists an orthogonal basis of* M_k^0 *made of eigenvectors of* $T(n)$ and that the eigenvalues of $T(n)$ are *real numbers.*

5.6.2. *Integrality properties.*

Let $M_k(\mathbf{Z})$ be the set of modular forms

$$f = \sum_{n=0}^{\infty} c(n)q^n$$

of weight $2k$ whose coefficients $c(n)$ are *integers.* One can prove that there exists a $\mathbf{Z}$-basis of $M_k(\mathbf{Z})$ which is a $\mathbf{C}$-basis of M_k. [More precisely, one can check that $M_k(\mathbf{Z})$ has the following basis (recall that $F = q\prod(1-q^n)^{24}$):

k even: One takes the monomials $E_2^{\alpha}F^{\beta}$ where $\alpha, \beta \in \mathbf{N}$, and $\alpha+3\beta = k/2$;

k odd: One takes the monomials $E_3E_2^{\alpha}F^{\beta}$ where α, $\beta \in \mathbf{N}$, and $\alpha+3\beta =$

$(k-3)/2$.] Proposition 12 shows that $M_k(\mathbf{Z})$ is stable under $T(n)$, $n \geqq 1$. We conclude from this that *the coefficients of the characteristic polynomial of* $T(n)$, acting on M_k, *are integers*[1]; in particular the eigenvalues of the $T(n)$ are *algebraic integers* ("totally real", by 5.6.1).

5.6.3. *The Ramanujan-Petersson conjecture.*

Let $f = \sum_{n \geqq 1} c(n)q^n$, $c(1) = 1$, be a cusp form of weight $2k$ which is a normalized eigenfunction of the $T(n)$.

Let $\Phi_{f,p}(T) = 1 - c(p)T + p^{2k-1}T^2$, p prime, be the polynomial defined in n° 5.4, formula (83). We can write

$$\Phi_{f,p}(T) = (1-\alpha_p T)(1-\alpha_p' T) \tag{89}$$

with

$$\alpha_p + \alpha_p' = c(p), \; \alpha_p \alpha_p' = p^{2k-1}. \tag{90}$$

The *Petersson conjecture* is that α_p and α_p' are *complex conjugate*. One can also express it by:

$$|\alpha_p| = |\alpha_p'| = p^{k-1/2},$$

or

$$|c(p)| \leqq 2p^{k-1/2},$$

or

$$|c(n)| \leqq n^{k-1/2}\sigma_0(n) \quad \text{for all } n \geqq 1.$$

For $k = 6$, this is the Ramanujan conjecture: $|\tau(p)| \leqq 2p^{11/2}$.

These conjectures have been proved in 1973 by P. Deligne (*Publ. Math. I.H.E.S.* n°43, p. 302), as consequences of the "Weil conjectures" about algebraic varieties over finite fields.

§6. *Theta functions*

6.1. *The Poisson formula*

Let V be a real vector space of finite dimension n endowed with an invariant measure μ. Let V' be the *dual* of V. Let f be a rapidly decreasing smooth function on V (see, L. SCHWARTZ, Théorie des Distributions, chap. VII, §3). The *Fourier transform* f' of f is defined by the formula

$$f'(y) = \int_V e^{-2i\pi\langle x,y\rangle} f(x)\mu(x). \tag{91}$$

This is a rapidly decreasing smooth function on V'.

Let now Γ be a *lattice* in V' (see n° 2.2). We denote by Γ' the lattice in V' *dual* to Γ; it is the set of $y \in V'$ such that $\langle x, y\rangle \in \mathbf{Z}$ for all $x \in \Gamma$. One

[1] We point out that there exists an *explicit formula* giving the trace of $T(n)$, cf. M. EICHLER and A. SELBERG, *Journ. Indian Math. Soc.*, 20, 1956.

checks easily that Γ' may be identified with the **Z**-dual of Γ (hence the terminology).

Proposition 15.—*Let* $v = \mu(V/\Gamma)$. *One has*:

$$\sum_{x\in\Gamma} f(x) = \frac{1}{v} \sum_{y\in\Gamma'} f'(y). \tag{92}$$

After replacing μ by $v^{-1}\mu$, we can assume that $\mu(V/\Gamma) = 1$. By taking a basis $e_1, \ldots, e_n$ of Γ, we identify V with $\mathbf{R}^n$, Γ with $\mathbf{Z}^n$, and μ with the product measure $dx_1 \ldots dx_n$. Thus we have $V' = \mathbf{R}^n$, $\Gamma' = \mathbf{Z}^n$ and we are reduced to the classical Poisson formula (SCHWARTZ, *loc. cit.*, formule (VII, 7:5)).

6.2. *Application to quadratic forms*

We suppose henceforth that V is endowed with a symmetric bilinear form $x.y$ which is *positive and nondegenerate* (i.e. $x.x > 0$ if $x \neq 0$). We identify V with V' by means of this bilinear form. The lattice Γ' becomes thus a *lattice* in V; one has $y \in \Gamma'$ if and only if $x.y \in \mathbf{Z}$ for all $x \in \Gamma$.

To a lattice Γ, we associate the following function defined on $\mathbf{R}_+^*$:

$$\Theta_\Gamma(t) = \sum_{x\in\Gamma} e^{-\pi t x.x}. \tag{93}$$

We choose the invariant measure μ on V such that, if $\varepsilon_1, \ldots \varepsilon_n$ is an orthonormal basis of V, the unit cube defined by the ε_i has volume 1. The volume v of the lattice Γ is then defined by $v = \mu(V/\Gamma)$, cf. n° 6.1.

Proposition 16.—*We have the identity*

$$\Theta_\Gamma(t) = t^{-n/2}v^{-1}\Theta_{\Gamma'}(t^{-1}). \tag{94}$$

Let $f = e^{-\pi x.x}$. It is a rapidly decreasing smooth function on V. The Fourier transform f' of f *is equal to* f. Indeed, choose an orthonormal basis of V and use this basis to identify V with $\mathbf{R}^n$; the measure μ becomes the measure $dx = dx_1 \ldots dx_n$ and the function f is

$$f = e^{-\pi(x_1^2 + \cdots + x_n^2)}.$$

We are thus reduced to showing that the Fourier transform of $e^{-\pi x^2}$ is $e^{-\pi x^2}$, which is well known.

We now apply prop. 15 to the function f and to the lattice $t^{1/2}\Gamma$; the volume of this lattice is $t^{n/2}v$ and its dual is $t^{-1/2}\Gamma'$; hence we get the formula to be proved.

6.3. *Matrix interpretation*

Let $e_1, \ldots, e_n$ be a basis of Γ. Put $a_{ij} = e_i.e_j$. The matrix $A = (a_{ij})$ is positive, nondegenerate and symmetric. If $x = \Sigma x_i e_i$ is an element of V, then

$$x.x = \Sigma a_{ij} x_i x_j.$$

The function Θ_Γ can be written

$$\Theta_\Gamma(t) = \sum_{x_i \in \mathbf{Z}} e^{-\pi t \Sigma a_{ij} x_i x_j}. \tag{95}$$

The *volume* v of Γ is given by:

$$v = \det(A)^{1/2}. \tag{96}$$

This can be seen as follows: Let $\varepsilon_1, \ldots, \varepsilon_n$ be an orthonormal basis of V and put

$$\varepsilon = \varepsilon_1 \wedge \ldots \wedge \varepsilon_n, \quad e = e_1 \wedge \ldots \wedge e_n.$$

We have $e = \lambda\varepsilon$ with $|\lambda| = v$. Moreover, $e.e = \det(A)\, \varepsilon.\varepsilon$, and by comparing, we obtain $v^2 = \det(A)$.

Let $B = (b_{ij})$ be the matrix inverse to A. One checks immediately that the dual basis (e_i') to (e_i) is given by the formulas:

$$e_i' = \sum b_{ij} e_j.$$

The (e_i') form a basis of Γ'. The matrix $(e_i'.e_j')$ is equal to B. This shows in particular that if $v' = \mu(V/\Gamma')$, then we have $vv' = 1$.

6.4. *Special case*

We will be interested in pairs (V, Γ) which have the following two properties:

(i) *The dual* Γ' *of* Γ *is equal to* Γ.

This amounts to saying that one has $x.y \in \mathbf{Z}$ for $x, y \in \Gamma$ and that the form $x.y$ defines an *isomorphism* of Γ onto its dual. In matrix terms, it means that the matrix $A = (e_i.e_j)$ has *integer coefficients* and that *its determinant equals* 1. By (96) the last condition is equivalent to $v = 1$.

If $n = \dim V$, this condition implies that the quadratic module Γ belongs to the category S_n defined in n° 1.1 of chap. V. Conversely, if $\Gamma \in S_n$ is positive definite, and if one puts $V = \Gamma \otimes \mathbf{R}$, the pair (V, Γ) satisfies (i).

(ii) *We have* $x.x \equiv 0 \pmod 2$ *for all* $x \in \Gamma$.

This means that Γ is *of type* II, in the sense of chap. V, n° 1.3.5, or else that the diagonal terms $e_i.e_i$ of the matrix A are *even*.

We have given in chap. V some examples of such lattices Γ.

6.5. *Theta functions*

In this section and the next one, we assume that the pair (V, Γ) satisfies conditions (i) and (ii) of the preceding section.

Let m be an integer ≥ 0, and denote by $r_\Gamma(m)$ the number of elements x of Γ such that $x.x = 2m$. It is easy to see that $r_\Gamma(m)$ is bounded by a polynomial in m (a crude volume argument gives for instance $r_\Gamma(m) = O(m^{n/2})$). This shows that the series with integer coefficients

$$\sum_{m=0}^{\infty} r_\Gamma(m)q^m = 1+r_\Gamma(1)q+\ldots$$

converges for $|q| < 1$. Thus one can define a function θ_Γ on the half plane H by the formula

(97)
$$\theta_\Gamma(z) = \sum_{m=0} r_\Gamma(m)q^m \quad (\text{where } q = e^{2\pi i z}).$$

We have:

(98)
$$\theta_\Gamma(z) = \sum_{x\in\Gamma} q^{(x.x)/2} = \sum_{x\in\Gamma} e^{\pi i z(x.x)}.$$

The function θ_Γ is called the *theta function* of the quadratic module Γ. It is holomorphic on H.

Theorem 8.—(a) *The dimension n of V is divisible by* 8.
(b) *The function θ_Γ is a modular form of weight $n/2$.*

Assertion (a) has already been proved (chap. V, n° 2.1, cor. 2 to th. 2). We prove the identity

(99)
$$\theta_\Gamma(-1/z) = (iz)^{n/2}\theta_\Gamma(z).$$

Since the two sides are analytic in z, it suffices to prove this formula when $z = it$ with t real >0. We have

$$\theta_\Gamma(it) = \sum_{x\in\Gamma} e^{-\pi t(x.x)} = \Theta_\Gamma(t).$$

Similarly, $\theta_\Gamma(-1/it) = \Theta_\Gamma(t^{-1})$. Formula (99) results thus from (94), taking into account that $v = 1$ and $\Gamma = \Gamma'$.

Since n is divisible by 8, we can rewrite (99) in the form

(100)
$$\theta_\Gamma(-1/z) = z^{n/2}\theta_\Gamma(z)$$

which shows that θ_Γ is a modular form of weight $n/2$.

[We indicate briefly another proof of (a). Suppose that n is not divisible by 8; replacing Γ, if necessary, by $\Gamma \oplus \Gamma$ or $\Gamma \oplus \Gamma \oplus \Gamma \oplus \Gamma$, we may suppose that $n \equiv 4 \pmod 8$. Formula (99) can then be written

$$\theta_\Gamma(-1/z) = (-1)^{n/4}z^{m/2}\theta_\Gamma(z) = -z^{n/2}\theta_\Gamma(z).$$

If we put $\omega(z) = \theta_\Gamma(z)dz^{n/4}$, we see that the differential form ω is transformed into $-\omega$ by $S:z \mapsto -1/z$. Since ω is invariant by $T:z \mapsto z+1$, we see that ST transforms ω into $-\omega$, which is absurd because $(ST)^3 = 1$.]

Corollary 1.—*There exists a cusp form f_Γ of weight $n/2$ such that*

(101)
$$\theta_\Gamma = E_k+f_\Gamma \quad \textit{where } k = n/4.$$

This follows from the fact that $\theta_\Gamma(\infty) = 1$, hence that $\theta_\Gamma - E_k$ is a cusp form.

Corollary 2.—*We have* $r_\Gamma(m) = \dfrac{4k}{B_k}\sigma_{2k-1}(m)+O(m^k)$ *where* $k = n/4$.

This follows from cor. 1, formula (34) and th. 5.

Remark.—The "error term" f_Γ is in general not zero. However Siegel has proved that the *weighted mean of the f_Γ is zero.* More precisely, let C_n be the set of classes (up to isomorphism) of lattices Γ verifying (i) and (ii) and denote by g_Γ the order of the automorphism group of $\Gamma \in C_n$ (cf. chap. V, n° 2.3). One has:

$$\sum_{\Gamma \in C_n} \frac{1}{g_\Gamma} \cdot f_\Gamma = 0 \tag{102}$$

or equivalently

$$\sum_{\Gamma \in C_n} \frac{1}{g_\Gamma} \theta_\Gamma = M_n E_k \quad \text{where } M_n = \sum_{\Gamma \in C_n} \frac{1}{g_\Gamma}. \tag{103}$$

Note that this is also equivalent to saying that the weighted mean of the θ_Γ is an *eigenfunction* of the $T(n)$.

For a proof of formulas (102) and (103), see C. L. SIEGEL, *Gesam. Abh.*, n° 20.

6.6. *Examples*

i) *The case $n = 8$.*

Every cusp form of weight $n/2 = 4$ is zero. Cor. 1 of th. 8 then shows that $\theta_\Gamma = E_2$, in other words:

$$r_\Gamma(m) = 240\sigma_3(m) \quad \text{for all integers } m \geqq 1. \tag{104}$$

This applies to the lattice Γ_8 constructed in chap. V, n° 1.4.3 (note that this lattice is the only element of C_8).

ii) *The case $n = 16$.*

For the same reason as above, we have:

$$\theta_\Gamma = E_4 = 1 + 480 \sum_{m=1}^{\infty} \sigma_7(m) q^m. \tag{105}$$

Here one may take $\Gamma = \Gamma_8 \oplus \Gamma_8$ or $\Gamma = \Gamma_{16}$ (with the notations of chap. V, n° 1.4.3); even though these two lattices are not isomorphic, they have the same theta function, i.e. they represent each integer the same number of times.

Note that the function θ attached to the lattice $\Gamma_8 \oplus \Gamma_8$ is the *square* of the function θ of Γ_8; we recover thus the identity:

$$\left(1 + 240 \sum_{m=1}^{\infty} \sigma_3(m) q^m\right)^2 = 1 + 480 \sum_{m=1}^{\infty} \sigma_7(m) q^m.$$

iii) *The case $n = 24$.*

The space of modular forms of weight 12 is of dimension 2. It has for basis the two functions:

$$E_6 = 1 + \frac{65520}{691} \sum_{m=1}^{\infty} \sigma_{11}(m) q^m,$$

$$F = (2\pi)^{-12}\Delta = q\prod_{m=1}^{\infty}(1-q^m)^{24} = \sum_{m=1}^{\infty}\tau(m)q^m.$$

The theta function associated with the lattice Γ can thus be written

(106) $$\theta_\Gamma = E_6 + c_\Gamma F \quad \text{with } c_\Gamma \in \mathbf{Q}.$$

We have

(107) $$r_\Gamma(m) = \frac{65520}{691}\sigma_{11}(m) + c_\Gamma\tau(m) \quad \text{for } m \geqq 1.$$

The coefficient c_Γ is determined by putting $m = 1$:

(108) $$c_\Gamma = r_\Gamma(1) - \frac{65520}{691}.$$

Note that it is $\neq 0$ since 65520/691 is not an integer.

Examples.

a) The lattice Γ constructed by J. LEECH (*Canad. J. Math.*, 16, 1964) is such that $r_\Gamma(1) = 0$. Hence:

$$c_\Gamma = -\frac{65520}{691} = -2^4.3^2.5.7.13/691.$$

b) For $\Gamma = \Gamma_8 \oplus \Gamma_8 \oplus \Gamma_8$, we have $r_\Gamma(1) = 3.240$, hence:

$$c_\Gamma = \frac{432000}{691} = 2^7 3^3 5^3/691.$$

c) For $\Gamma = \Gamma_{24}$, we have $r_\Gamma(1) = 2.24.23$, hence:

$$c_\Gamma = \frac{697344}{691} = 2^{10}3.227/691.$$

6.7. *Complements*

The fact that we consider only the full modular group $G = \mathbf{PSL}_2(\mathbf{Z})$, forced us to limit ourselves to lattices verifying the very restrictive conditions of n° 6.4. In particular, we have not treated the most natural case, that of the quadratic forms

$$x_1^2 + \ldots + x_n^2,$$

which verify (i) but not (ii). The corresponding theta functions are "modular forms of weight $n/2$" (note that $n/2$ is not necessarily an integer) with respect to the subgroup of G generated by S and T^2. This group has index 3 in G, and its fundamental domain has two "cusps" to which correspond two types of "Eisenstein series"; using them, one obtains formulas giving the number of representations of an integer as a sum of n squares; for more details, see the books and papers quoted in the bibliography.

Bibliography

Some classics

C. F. Gauss—*Disquisitiones arithmeticae*, 1801, *Werke*, Bd. I. (English translation: Yale Univ. Press—French translation: Blanchard.)
C. Jacobi—*Fundamenta nova theoriae functionum ellipticarum*, 1829, *Gesammelte Werke*, Bd. I., pp. 49–239.
G. Lejeune Dirichlet—*Démonstration d'un théorème sur la progression arithmétique*, 1834, *Werke*, Bd. I, p. 307.
G. Eisenstein, *Mathematische Werke*, Chelsea, 1975.
B. Riemann—*Gesammelte mathematische Werke*, Teubner-Springer-Verlag, 1990 (English translation: Dover—partial French translation: Gauthier-Villars, 1898).
D. Hilbert—*Die Theorie der algebraischer Zahlkörper*, *Gesam. Abh.*, Bd. I, pp. 63–363 (French translation: *Ann. Fac. Sci. Toulouse*, 1909 and 1910).
H. Minkowski—*Gesammelte Abhandlungen*, Teubner, 1911; Chelsea, 1967.
A. Hurwitz—*Mathematische Werke*, Birkhäuser Verlag, 1932.
E. Hecke—*Mathematische Werke*, Göttingen, 1959.
C. L. Siegel—*Gesammelte Abhandlungen*, Springer-Verlag, 1966–1979.
A. Weil—*Collected Papers*, Springer-Verlag, 1980.

Number fields and local fields

E. Hecke—*Algebraische Zahlen*, Leipzig, 1923.
Z. I. Borevich and I. R. Shafarevich—*Number Theory* (translated from Russian) Academic Press, 1966. (There exist also translations into French and German.)
M. Eichler—*Einführung in die Theorie der algebraischen Zahlen und Funktionen*, Birkhäuser Verlag, 1963 (English translation: Academic Press, 1966).
J-P. Serre—*Corps Locaux*, Hermann, 1962.
P. Samuel—*Théorie algébrique des nombres*, Hermann, 1967.
E. Artin and J. Tate—*Class Field Theory*, Benjamin, 1968.
J. Cassels and A. Fröhlich (edit.)—*Algebraic Number Theory*, Academic Press, 1967.
A. Weil—*Basic Number Theory*, Springer-Verlag, 1967.
S. Lang—*Algebraic Number Theory*, Addison-Wesley, 1970.
(The four last works contain an exposition of the so-called "class field theory".)

Quadratic forms

a) *Generalities, Witt's theorem*

E. Witt—*Theorie der quadratischen Formen in beliebigen Körpern*, J. Crelle, 176, 1937, pp. 31–44.
N. Bourbaki—*Algèbre*, chap. IX, Hermann, 1959.
E. Artin—*Geometric Algebra*, Interscience Publ., 1957 (French translation: Gauthier-Villars, 1962).

b) *Arithmetic properties*

B. Jones—*The arithmetic theory of quadratic forms*, Carus Mon., n° 10, John Wiley and Sons, 1950.
M. Eichler—*Quadratische Formen und orthogonale Gruppen*, Springer-Verlag, 1952.
G. L. Watson—*Integral quadratic forms*, Cambridge Tracts, n° 51. Cambridge, 1960.
O. T. O'Meara—*Introduction to quadratic forms*. Springer-Verlag, 1963.
J. Milnor and D. Husemoller—*Symmetric Bilinear Forms*, Springer-Verlag, 1973.
T. Y. Lam—*The algebraic theory of quadratic forms*, New York, Benjamin, 1973.
J. W. S. Cassels—*Rational Quadratic Forms*, Academic Press, 1978.

c) *Integral quadratic forms with discriminant* ± 1

E. WITT—*Eine Identität zwischen Modulformen zweiten Grades*, Abh. math. Sem. Univ. Hamburg, 14, 1941, pp. 323–337.
M. KNESER—*Klassenzahlen definiter quadratischer Formen*, Arch. der Math. 8, 1957, pp. 241–250.
J. MILNOR—*On simply connected manifolds*, Symp. Mexico, 1958, pp. 122–128.
J. MILNOR—*A procedure for killing homotopy groups of differentiable manifolds*, Symp. Amer. Math. Soc., n° 3, 1961, pp. 39–55.
J. H. CONWAY and N. J. A. SLOANE—*Sphere Packings, Lattices and Groups*, Springer-Verlag, 1988.

Dirichlet theorem, zeta function and L-functions

J. HADAMARD—*Sur la distribution des zéros de la fonction $\zeta(s)$ et ses conséquences arithmétiques*, 1896, Oeuvres, CNRS, t. 1, pp. 189–210.
E. LANDAU—*Handbuch der Lehre von der Verteilung der Primzahlen*, Teubner, 1909; Chelsea, 1953.
A. SELBERG—*An elementary proof of the prime number theorem for arithmetic progressions*, Canad. J. Math., 2, 1950, pp. 66–78.
E. C. TITCHMARSH—*The Theory of the Riemann zeta-function*, Oxford, 1951.
K. PRACHAR—*Primzahlverteilung*, Springer-Verlag, 1957.
H. DAVENPORT—*Multiplicative number theory*, second edition, Springer-Verlag, 1980.
K. CHANDRASEKHARAN—*Introduction to analytic number theory*, Springer-Verlag, 1968.
A. BLANCHARD—*Initiation à la théorie analytique des nombres premiers*, Dunod, 1969.
H. M. EDWARDS—*Riemann's zeta function*, New York, Acad. Press, 1974.
W. NARKIEWICZ—*Elementary and analytic theory of algebraic numbers*, Warsaw, Mon. Mat. 57, 1974.
W. ELLISON—*Les Nombres Premiers*, Paris, Hermann, 1975.

Modular functions

F. KLEIN—*Vorlesungen über die Theorie der elliptischen Modulfunktionen*, Leipzig, 1890.
S. RAMANUJAN—*On certain arithmetical functions*, Trans. Cambridge Phil. Soc., 22, 1916, pp. 159–184 (= *Collected Papers*, pp. 136–162).
G. HARDY—*Ramanujan*, Cambridge Univ. Press, 1940.
R. GODEMENT—*Travaux de Hecke*, Sém. Bourbaki, 1952–53, exposés 74, 80.
R. C. GUNNING—*Lectures on modular forms* (notes by A. Brumer), Ann. of Math. Studies, Princeton, 1962.
A. BOREL et al.—*Seminar on complex multiplication*, Lecture Notes in Maths., n° 21, Springer-Verlag, 1966.
A. OGG—*Modular forms and Dirichlet series*, Benjamin, 1969.
G. SHIMURA—*Introduction to the arithmetic theory of automorphic functions*, Tokyo–Princeton, 1971.
H. RADEMACHER—*Topics in Analytic Number Theory*, Springer-Verlag, 1973.
W. KUYK et al. (edit.)—*Modular Functions of One Variable*, I, . . . , VI, *Lecture Notes in Math.*, 320, 349, 350, 476, 601, 627, Springer-Verlag, 1973–1977.
P. DELIGNE—*La conjecture de Weil* I, Publ. Math. I.H.E.S., 43, 1974, p. 273–307.
A. WEIL—*Elliptic Functions according to Eisenstein and Kronecker*, Springer-Verlag, 1976.
S. LANG—*Introduction to Modular Forms*, Springer-Verlag, 1976.
R. RANKIN—*Modular Forms and Functions*, Cambridge Univ. Press, 1977.

(See also the works of HECKE, SIEGEL and WEIL quoted above.)

Index of Definitions

Index of Notations

Graduate Texts in Mathematics

(continued from page ii)

133 HARRIS. Algebraic Geometry: A First Course.
134 ROMAN. Coding and Information Theory.
135 ROMAN. Advanced Linear Algebra.
136 ADKINS/WEINTRAUB. Algebra: An Approach via Module Theory.
137 AXLER/BOURDON/RAMEY. Harmonic Function Theory.
138 COHEN. A Course in Computational Algebraic Number Theory.
139 BREDON. Topology and Geometry.
140 AUBIN. Optima and Equilibria. An Introduction to Nonlinear Analysis.
141 BECKER/WEISPFENNING/KREDEL. Gröbner Bases. A Computational Approach to Commutative Algebra.
142 LANG. Real and Functional Analysis. 3rd ed.
143 DOOB. Measure Theory.
144 DENNIS/FARB. Noncommutative Algebra.
145 VICK. Homology Theory. An Introduction to Algebraic Topology. 2nd ed.
146 BRIDGES. Computability: A Mathematical Sketchbook.
147 ROSENBERG. Algebraic *K*-Theory and Its Applications.
148 ROTMAN. An Introduction to the Theory of Groups. 4th ed.
149 RATCLIFFE. Foundations of Hyperbolic Manifolds.
150 EISENBUD. Commutative Algebra with a View Toward Algebraic Geometry.
151 SILVERMAN. Advanced Topics in the Arithmetic of Elliptic Curves.
152 ZIEGLER. Lectures on Polytopes.
153 FULTON. Algebraic Topology: A First Course.
154 BROWN/PEARCY. An Introduction to Analysis.
155 KASSEL. Quantum Groups.
156 KECHRIS. Classical Descriptive Set Theory.
157 MALLIAVIN. Integration and Probability.
158 ROMAN. Field Theory.
159 CONWAY. Functions of One Complex Variable II.
160 LANG. Differential and Riemannian Manifolds.
161 BORWEIN/ERDÉLYI. Polynomials and Polynomial Inequalities.
162 ALPERIN/BELL. Groups and Representations.
163 DIXON/MORTIMER. Permutation Groups.
164 NATHANSON. Additive Number Theory: The Classical Bases.
165 NATHANSON. Additive Number Theory: Inverse Problems and the Geometry of Sumsets.
166 SHARPE. Differential Geometry: Cartan's Generalization of Klein's Erlangen Program.
167 MORANDI. Field and Galois Theory.
168 EWALD. Combinatorial Convexity and Algebraic Geometry.
169 BHATIA. Matrix Analysis.
170 BREDON. Sheaf Theory. 2nd ed.
171 PETERSEN. Riemannian Geometry.
172 REMMERT. Classical Topics in Complex Function Theory.
173 DIESTEL. Graph Theory. 2nd ed.
174 BRIDGES. Foundations of Real and Abstract Analysis.
175 LICKORISH. An Introduction to Knot Theory.
176 LEE. Riemannian Manifolds.
177 NEWMAN. Analytic Number Theory.
178 CLARKE/LEDYAEV/STERN/WOLENSKI. Nonsmooth Analysis and Control Theory.
179 DOUGLAS. Banach Algebra Techniques in Operator Theory. 2nd ed.
180 SRIVASTAVA. A Course on Borel Sets.
181 KRESS. Numerical Analysis.
182 WALTER. Ordinary Differential Equations.
183 MEGGINSON. An Introduction to Banach Space Theory.
184 BOLLOBAS. Modern Graph Theory.
185 COX/LITTLE/O'SHEA. Using Algebraic Geometry.
186 RAMAKRISHNAN/VALENZA. Fourier Analysis on Number Fields.
187 HARRIS/MORRISON. Moduli of Curves.
188 GOLDBLATT. Lectures on the Hyperreals: An Introduction to Nonstandard Analysis.
189 LAM. Lectures on Modules and Rings.
190 ESMONDE/MURTY. Problems in Algebraic Number Theory.
191 LANG. Fundamentals of Differential Geometry.
192 HIRSCH/LACOMBE. Elements of Functional Analysis.
193 COHEN. Advanced Topics in Computational Number Theory.
194 ENGEL/NAGEL. One-Parameter Semigroups for Linear Evolution Equations.
195 NATHANSON. Elementary Methods in Number Theory.
196 OSBORNE. Basic Homological Algebra.
197 EISENBUD/HARRIS. The Geometry of Schemes.
198 ROBERT. A Course in *p*-adic Analysis.
199 HEDENMALM/KORENBLUM/ZHU. Theory of Bergman Spaces.

200 BAO/CHERN/SHEN. An Introduction to Riemann–Finsler Geometry.
201 HINDRY/SILVERMAN. Diophantine Geometry: An Introduction.
202 LEE. Introduction to Topological Manifolds.
203 SAGAN. The Symmetric Group: Representations, Combinatorial Algorithms, and Symmetric Function. 2nd ed.
204 ESCOFIER. Galois Theory.
205 FÉLIX/HALPERIN/THOMAS. Rational Homotopy Theory.

Printed in the United States
By Bookmasters